T0320654

CRYSTAL GROWTH

Principles and Progress

UPDATES IN APPLIED PHYSICS AND ELECTRICAL TECHNOLOGY

Series Editor: P. J. Dobson
Philips Research Laboratories
Redhill, England

PHOTODETECTORS: An Introduction to Current Technology
P. N. J. Dennis

CRYSTAL GROWTH: Principles and Progress
A. W. Vere

A Continuation Order Plan is available for this series. A continuation order will bring delivery of each new volume immediately upon publication. Volumes are billed only upon actual shipment. For further information please contact the publisher.

CRYSTAL GROWTH

Principles and Progress

A. W. Vere

Royal Signals and Radar Establishment
Great Malvern, England

Plenum Press • New York and London

Library of Congress Cataloging in Publication data

Vere, A. W.
　Crystal growth.
　(Updates in applied physics and electrical technology)
　Bibliography: p.
　Includes index.
　1. Crystals–Growth. I. Title. II. Series.
QD921.V44　1987　　　　　　　548'.5　　　　　　　87-7751
ISBN 0-306-42576-9

© 1987 Plenum Press, New York
A Division of Plenum Publishing Corporation
233 Spring Street, New York, N.Y. 10013

Text, excluding abstracts,
© 1987 Her Majesty's Stationery Office

Printed in the United States of America

To Penny, Bernard, and Tim
in appreciation of their continued encouragement

PREFACE

This book is the second in a series of scientific textbooks designed to cover advances in selected research fields from a basic and general viewpoint, so that only limited knowledge is required to understand the significance of recent developments. Further assistance for the non-specialist is provided by the summary of abstracts in Part 2, which includes many of the major papers published in the research field.

Crystal Growth of Semiconductor Materials has been the subject of numerous books and reviews and the fundamental principles are now well-established. We are concerned chiefly with the deposition of atoms onto a suitable surface - crystal growth - and the generation of faults in the atomic structure during growth and subsequent cooling to room temperature - crystal defect structure. In this book I have attempted to show that whilst the fundamentals of these processes are relatively simple, the complexities of the interactions involved and the individuality of different materials systems and growth processes have ensured that experimentally verifiable predictions from scientific principles have met with only limited success - good crystal growth remains an art. However, recent advances, which include the reduction of growth temperatures, the reduction or elimination of reactant transport variables and the use of better-controlled energy sources to promote specific reactions, are leading to simplified growth systems. This progress, coupled with the increasing development and application of *in situ* diagnostic techniques to monitor, and perhaps ultimately control, the crystal-growth process, is now rapidly transforming the field from an art to a science. It is also incidentally proving the correctness of many early theories whose experimental verification had not previously been possible.

The aims of this book are, therefore, to chart the progress of these developments and to illustrate the way in which the field is developing, in the hope that it will provide a stimulus to thought for those already involved in the field, but will also serve as a general introduction to the subject for chemists, surface scientists, high vacuum technologists and followers of many other disciplines whose contributions are increasingly required.

ACKNOWLEDGEMENTS

It is a pleasure to acknowledge the generous help and support given to me in the preparation of this book by all my colleagues in the semiconductor materials growth divisions at RSRE. In particular, I would like to thank Dr Don ' Hurle and Dr Dennis Williams for their editing of the text and their many useful comments and suggestions. Nigel Chew, Colin Warwick and Graham Brown assisted in the compilation of Chapter 3 and provided many of the illustrations for that Chapter, whilst Brian Cockayne and Mike Astles contributed similarly to Chapter 4. Chapter 5 was compiled with assistance from Colin Whitehouse and Stuart Irvine.

In seeking to present as broad a picture of the subject as possible, I have necessarily drawn upon the work and experience of semiconductor materials scientists throughout the community. Many have not only given permission for the reproduction of their work but have contributed supplementary material and constructive criticism. Their work (and the consent of their publishers for the reproduction of the material) is acknowledged at the appropriate point in the text. I would also like to thank the Ministry of Defence for permission to reproduce Figs. 3.10, 11, 12, 13, 17, 18, 21, 22, 24, 26, 27, Figs. 4.8, 14, 16, and Fig. 5.13.

Finally, but by no means least, my thanks to Peggy Cox for all her hard work on the typing and editing of the manuscript and for her patience and perpetual cheerfulness, despite my many revisions of the text.

A W Vere
Royal Signals and Radar Establishment
Malvern
1986

CONTENTS

CHAPTER 1

INTRODUCTION

When man first started to grow crystals he used the ways already perfected in nature, suspending small 'seed' crystals of the required material within a supersaturated solution or flux. Although frequently beautiful, the resulting crystals were small and imperfect, containing inclusions, impurities and stacking faults in the crystal lattice.

To produce larger crystals, researchers like Czochralski (1917), Bridgman (1925), Kyropoulus (1926) and Stockbarger (1936) invented 'bulk-growth' techniques in which the required material was slowly cooled through its melting point, either by moving its container through a steep temperature gradient, or by slow, controlled, removal of a seed crystal from the melt. These techniques have been increasingly refined, to the extent that silicon semiconductor material can now be pulled by the Czochralski technique in the form of single-crystals eight inches in diameter and several feet long. Even today, however, the high growth temperatures involved lead to impurity pick-up from the crucible and other growth equipment, while the high thermal gradients required to stabilise the liquid/solid growth interface result in internal strain and the production of crystal lattice point and line defects which degrade the performance of electronic devices.

It is one of the myths of semiconductor technology that "epitaxial" growth techniques, in which thin layers of the required material are deposited on a suitably oriented crystalline substrate, have largely superseded bulk growth for electronic circuit fabrication because the devices to be fabricated require only micron dimensions. The use of epitaxial growth, therefore, reduces growth times and wafering costs and eliminates waste due to ingot slicing 'kerf' losses. In fact, the cost and time of preparation of the bulk substrate must be added to those of the epitaxial growth process. There are, however, three major advantages. Firstly, crystal growth is, by definition, a non-equilibrium reaction and the probability of defect generation increases with growth time. Thin-layer growth is therefore possible under conditions which, if permitted to continue, would inevitably lead to unacceptable defect levels. Secondly, because the layers are grown from vapour, flux or solution, a lower growth temperature is possible, so reducing contamination. Finally, exposure to different vapours or liquids can be used to produce multilayer growth.

Although liquid phase epitaxy, LPE, has achieved many notable successes, significant problems exist in growing multilayer structures by the technique due to incomplete 'wipe-off' of the liquid from the previous layer deposition and to variable depth "melt-back" frequently encountered during growth on imperfect substrates. Vapour phase epitaxy, VPE, is increasingly preferred except where suitable high vapour-pressure precursors are not available (eg oxide and garnet growth). Originally LPE growth was used in preference to VPE because the high temperatures required to produce vapour-phase species from elemental or alloy sources led to transport-limited growth, with attendant problems of non-uniform mass-transport due to convection in the vapour phase and diffusion through the boundary layer adjacent to the substrate. The situation was reversed by the development of new, lower temperature vapour-phase techniques such as metalorganic vapour phase epitaxy, MOVPE (e.g. Manasevit, 1968; Bass, 1975) and molecular beam epitaxy, MBE (e.g. Arthur, 1968; Joyce, 1974; Cho and Arthur, 1975). These techniques transferred the emphasis from mass-transport and diffusion controlled growth to surface adsorption and migration controlled growth. At the same time, particularly in the case of MBE, they provided the diagnostic techniques, such as mass spectro-metry, electron energy loss (EELs), reflection high energy electron diffraction (RHEED) and others, to study such surface reactions. In the next chapter I have attempted to show, from a brief historical survey, the development of nucleation and growth theories and the way in which many aspects of early theories, which could not be proven in the relatively crude experimental growth processes avail-able at the time of their conception, are now capable of modification and extension in the light of practical knowledge derived from the modern vapour-phase techniques.

Nevertheless, many crystal defect types and sources still exist and the trend to lower temperature crystal growth, whilst reducing the stress and diffusion-induced defects characteristic of growth from the melt, has introduced (through reduced atomic mobility on the surface and reduced desorption rates) a whole new class of layer/substrate interface defects. Although several books and reviews have dealt extensively with the subject of defects, in Chapter 3 I have emphasised the similarity between defects in different semiconductor materials and introduced some of the more recent work on interfacial defects.

Since bulk-growth remains important for devices such as solid state lasers and optoelectronic components, but also more especially for large area, crystalline substrates for epitaxy, Chapter 4 reviews recent developments in this field. These include the development of computer models of thermal fields within the crystal and melt and experimental techniques to reduce the effects of convection on crystal non-uniformity.

The accent throughout this volume is on the advantages of low-temperature growth. It is therefore necessary to consider in some detail the surface kinetic limitations which govern the lowest deposition temperature attainable. For this reason the review of chemical vapour epitaxy (CVE) techniques in Chapter 5 emphasises the trend towards kinetically-limited growth and the consequent increase in importance of low-pressure and plasma-assisted techniques, whilst the treatment of MBE considers the recent advances in our knowledge of substrate surfaces and adsorption/desorption reactions resulting from UHV deposition.

The final chapter explores some of the new structures which have been made possible by modern vapour growth techniques. Perhaps the most spectacular advances have been in the field of superlattices and multiple quantum-well structures comprising hundreds of alternating layers of differing compositions, individual layers being only a few tens of Angstroms thick. The physics of such devices is completely new and offers exciting prospects in terms of faster, even smaller electronic components. We examine, too, the latest developments in CVE techniques aimed at computer-controlled, selected-area deposition and etching of very thin multilayer structures, leading to the capability to perform wafer-scale integration within the same growth equipment. At the same time, the increased sophistication of growth control and in situ real-time analysis of the growth process is already leading to a better understanding of substrate surface damage and reconstruction, the nature and kinetics of adsorption and desorption species and the nucleation and growth mechanisms. Finally, in a brief introduction to the growing subject of photon-assisted deposition, we show how, in addition to the technologically important features such as high-resolution selected-area growth and growth rate control through optical modulation, selection of the frequency and amplitude of the photon source can be used to promote specific precursor reactions, leading to optimisation of the surface adsorption and reaction-product desorption. These advances bring us a step nearer to the materials scientists's ultimate goal of being able to select the optimum precursor species, control their

adsorption and migration on a well-prepared substrate surface and to
arrange for the efficient removal of unwanted reaction products.

CHAPTER 2

TRANSPORT, NUCLEATION AND GROWTH

2.1 Introduction

Most crystal growth processes involve the following steps:

1. Generation of reactants

2. Transport of reactants to the growth surface (in some cases via a "boundary layer" adjacent to the surface).

3. Adsorption at the growth surface

4. Nucleation (irreversible location on a crystal lattice site)

5. Growth (advance of the liquid-solid or vapour-solid interface)

6. Removal of unwanted reaction products from the growth surface

These are illustrated schematically in Figure 2.1. The driving force for crystallisation is the supersaturation of the gas or liquid phase with respect to the component whose growth is required. Too little supersaturation will result in an unacceptably slow growth rate. At the other extreme of the supersaturation range the rate of condensation exceeds the rate at which the atoms or molecules can be incorporated onto the crystal lattice, leading to breakdown of the single-crystal interface and the onset of non-uniform cellular or dendritic growth (see Chapter 3). The object of good crystal growth is therefore to achieve and maintain a constant level of supersaturation within this range.

VAPOUR PHASE GROWTH

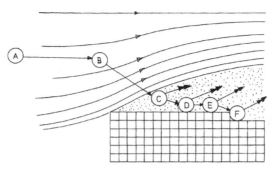

LIQUID PHASE GROWTH

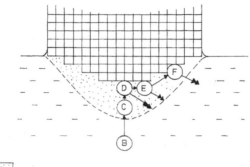

Boundary Layer   Waste Product ➤➤

Fig.2.1   Basic steps in crystal growth from the vapour or
           liquid phase. A. Generation.   B. Bulk Transport.
           C. Boundary layer transport.   D. Adsorption/desorption.
           E. Migration.   F. Nucleation

If the linear growth rate of an advancing gas/solid interface is
plotted as a function of flow rate, the generalised form of the
relationship is frequently as shown in Figure 2.2. In the low
flow-rate regime the growth is determined by the rate of arrival of
reactants at the growth interface and is approximately proportional
to flow rate. In the high flow rate regime a sufficient supersaturation of solute is always available, but the growth rate is
dependent upon the rate of chemical reaction and incorporation at
the growth surface - it is said to be kinetically limited.

    It is perhaps surprising to find a transport-limited regime in
the case of liquid-phase growth where solute and growth interface
are in intimate contact. However, in both gas-phase and liquid-phase

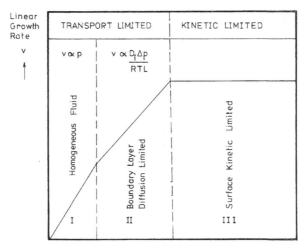

Fig.2.2   Idealised growth-rate versus fluid flow-rate plot showing
the different growth regimes

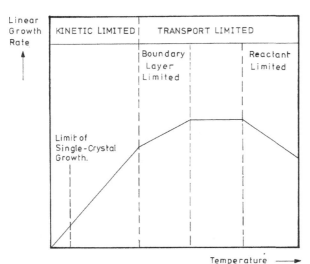

Fig.2.3   Idealised plot of growth rate versus deposition
temperature showing the different growth regimes
(Boundary layer formation and extent are dependent upon
temperature gradient above the growth surface)

systems the supersaturation at the growth surface may be locally
reduced by the formation of a boundary layer of different
concentration to the bulk liquid or gas. This boundary layer may be
due either to the rejection of solute ahead of the interface in
solution or melt growth (a solutal boundary) or to a flow rate
gradient (a hydrodynamic boundary layer). Forced or convective flow
frequently leads to reduction of the solutal boundary layer
thickness, so that in this case the boundary layer thickness and
concentrations are dependent upon both the hydrodynamics of the
reaction vessel and the segregation properties of the material. We
may therefore sub-divide the transport-limited growth regime of
Figure 2.2 into a homogeneous fluid regime and a boundary-layer
diffusion regime.

    In a similar manner Figure 2.3 shows schematically the
transition from surface/kinetic- to transport-limited growth
occurring in vapour phase deposition as the growth temperature
increases. At low temperatures the growth of single crystal
material is difficult because the adsorption rate exceeds the rate
of migration to suitable nucleation sites and the reaction product
desorption rate is also low. At high temperatures the growth rate
may begin to fall in the transport-limited regime due to limitations
in the reactant supply rate. The more likely situation, however, is
that the reaction becomes thermodynamically-controlled and the fall
in growth rate at high temperatures is more frequently attributable
to the exothermic nature of the reaction.

2.2   Generation of Reactants

    Vapour phase reactants may be generated in a variety of ways.
The simplest of these is the heating of volatile components to cause
melting or sublimation, eg.

$$Cd_{(s)} \rightarrow Cd_{(g)} \qquad\qquad (2.1)$$

$$Te_{(s)} \rightarrow Te_{(g)} \qquad\qquad (2.2)$$

$$Cd_{(g)} + Te_{(g)} \rightarrow CdTe_{(s)} \qquad\qquad (2.3)$$

    Alternatively the required reactants may be generated from the
pyrolytic (or photolytic) decomposition of a hydride, metalorganic
or other vapour-phase compound, eg:

$$SiH_4 \rightarrow Si + 2H_2 \qquad\qquad (2.4)$$

$$(CH_3)_2Cd+(C_2H_5)_2Te \rightarrow CdTe + (CH_4,C_2H_6,etc.) \qquad\qquad (2.5)$$

In other cases the required reactant, particularly the metallic component, eg Ga or In in GaAs or InP, is chemically reacted with a volatile gas to produce a meta-stable transport product, eg:

$$2Ga + 3Cl_2 \leftrightarrows 2GaCl_3 \qquad (2.6)$$

$$GaCl_3 + AsH_3 \leftrightarrows GaAs + 3HCl \qquad (2.7)$$

Cl_2 and I are frequently used as transporting agents. Equations (2.1) - (2.3) involve only the chemical species required for growth of the end-product and are therefore classed as vapour transport reactions, whereas equations (2.4) - (2.7) involve other reactants in chemical combinations with the required growth components and are therefore classed as Chemical Vapour Deposition (CVD) reactions.

2.3 Transport of Reactants

2.3.1 Bulk vapour transport

If we assume that the vapour phase environment between the source and substrate is a homogeneous medium in which the only independent variable is the temperature difference between source and substrate and the only dependent variables are p_A, p_B, p_C, p_D, the partial pressures of the reactants and products in the reaction

$$A + B \leftrightarrows C + D \qquad (2.8)$$

then the resulting reaction rate (and hence growth rate) can be determined.

We can illustrate this by considering the analysis of Theeten et al (1976). The deposition of GaAs occurs through the reaction

$$GaAs_{(s)} + HCl \leftrightarrows GaCl + 1/4\ As_4 + 1/2\ H_2 \qquad (2.9)$$

in which the forward reaction occurs at the source ($T_s = 850°C$) whilst the reverse reaction leads to deposition at the substrate ($T_D \simeq 750°C$). The corresponding reaction rate constant is given by

$$K(T_s) = P_{GaCl} \cdot P_{As_4}^{1/4} \cdot P_{H_2}^{1/2} \cdot P_{HCl}^{-1} \qquad (2.10)$$

In practice, transport between source and substrate is achieved using H_2 as a carrier gas so that the corresponding partial pressure p' of each component on arrival at the substrate is then given by

$$P'_{As_4} = P_{As_4} \cdot d \qquad (2.11)$$

$$P'_{GaCl} = P_{GaCl} \cdot d \qquad (2.12)$$

$$P'_{HCl} = P_{HCl} \cdot d \qquad (2.13)$$

where d is the dilution factor due to the hydrogen carrier gas pressure. The supersaturation α at the substrate is then given by

$$\alpha = \frac{(p'_{GaCl})(p'_{As_4})^{1/4}(p'_{H_2})^{1/2}(p'_{HCl})^{-1}}{K(T_D)} \qquad (2.14)$$

where $K(T_D)$ is the rate constant for the reverse reaction (2.9) at the substrate temperature T_D. Substituting (2.10) into (2.14) we obtain

$$\alpha = \frac{d^{1/4} K(T_S)}{K(T_D)} \qquad (2.15)$$

so we see that, in the transport-limited regime the supersaturation and hence the growth rate are directly proportional to the source/substrate temperature ratio. Although the result has been experimentally demonstrated by several authors (see, for example, Shaw 1968), the temperature range over which T_D may be varied whilst still affording single-crystal growth is small (700-800°C). Decreasing T_S leads to the expected transition from transport-limited to surface-kinetically limited growth for T_S >700°C whilst increasing T_S above 800°C leads to a reduced reaction rate through reduced surface lifetime of the adsorbed species. Table 1 shows the approximate growth temperature ranges for some common semiconductor materials. Note that in metalorganic growth systems the low temperature limit is frequently set by the minimum temperature required for efficient pyrolysis of the organic precursor. In LPE the obvious lowest limit is the freezing point of the flux or solution. It should be emphasised that the temperature ranges shown are approximations. The lower limit in particular depends critically on substrate preparation. In many cases higher temperature preparation is an essential prerequisite of low temperature growth.

Finally, it should be noted that in practical vapour growth systems the linear dependence of growth rate on *flow rate* is not always observed. Non-linearity arises from changes in the gas stream conditions, which tend to deplete the reactant concentration with increasing flow, giving rise to sub-linear conditions. These include the formation of increasingly broad boundary layers over the substrate surface, the onset of turbulence in the gas stream and an increase of homogeneous gas phase nucleation at localised cold spots away from the substrate.

Table 2.1   Growth temperature ranges for some common
             semiconductor materials

Material	Technique	Temperature °C		
Si	Bulk			1414
	LPCVD	1100	-	600
	MBE	800	-	500
GaAs	Bulk			1238
	MOVPE	550	-	850
	MBE	550	-	650
	VPE (HCl)	700	-	800
InP	Bulk			1062
	MOVPE	550	-	750
	MBE	400	-	500
	VPE	630	-	700
$Cd_{0.2}Hg_{0.8}Te$	Bulk			690
	LPE	450	-	550
	MOVPE	410	-	450
	MBE	180	-	220
ZnS	MOVPE	250	-	550

2.3.2 Boundary-layer transport

In vapour-phase growth at moderate flow rates the imposition of
a susceptor and single-crystal substrate in the gas stream causes a
perturbation in the flow, as shown in Figure 2.1(a). The shaded
area represents a region of relatively slow-moving gas with the flow
rate increasing from zero at the gas/substrate interface to the bulk
stream value at the edge of the boundary layer. The existence of
such layers was first observed by Bunn (1949). The treatment of
mass transport across them has been reviewed by Shaw (1974).

For the simplest case of mass transport by molecular diffusion
of a species i with diffusion coefficient D_i across a boundary layer
of constant thickness L the mass transfer coefficient k_g is given by
Hougen and Watson (1947)

$$k_g = \frac{D_i \Delta P}{RTL} \qquad (2.16)$$

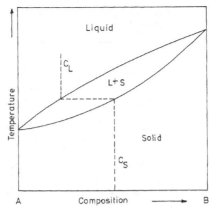

Fig.2.4   Idealised form of the phase diagram for a binary solid
          solution. On cooling, liquid of composition C_L
          precipitates solid of composition C_S.

where P is the pressure difference p_o-p_{eq} (p_o = bulk pressure p_{eq} =
interface pressure, assumed to be the equilibrium pressure). In
practical systems, however, the layer thickness varies with position
and the concentration of the depositing species. In the simpler
approximations this is assumed to vary linearly from zero at the
deposition surface to the bulk stream value at the outer edge of the
boundary layer. In reality it is subject to local variations
resulting from irregularities in the gas flow. Additional
complications are introduced by the need to consider multicomponent
diffusion to take into account interaction between the reactant (and
product) species. Nevertheless the simple approximation shown in
equation (2.16) is valid for many systems. If we assume that the
linear growth rate is proportional to the rate of mass transport
across the boundary layer, we see that the growth rate is again
linearly related to p_o and hence system flow rate and is inversely
proportional to the boundary layer thickness.

Figure 2.4 shows the idealised phase diagram for a binary solid
solution. To precipitate solid of composition C_S it is necessary to
start from a melt composition of C_L. In the system illustrated the
equilibrium segregation coefficient

$$k = \frac{C_S}{C_L}$$
                                                                      (2.17)

is large ($k \simeq 2$). As the alloy solidifies, atoms are preferentially
incorporated at the interface leading to a build up of A atom
concentration in the liquid ahead of the interface. The concen-

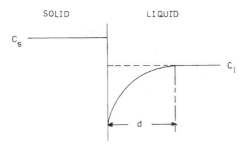

Fig.2.5   Schematic diagram of the concentration profile of an alloy
           component or impurity distribution across the solid/liquid
           interface, showing the formation of a depletion boundary
           layer in the liquid phase (k>1)

tration profile is then schematically as shown in Figure 2.5,
resulting in the formation of an A-component-depleted boundary
layer. Similar depletion (or, for k<1, enhancement) boundary layers
are observed in many Czochralski and solution growth systems (see
for example Elwell and Scheel, 1975). In the absence of a boundary
layer the linear growth rate v is proportional to the rate of mass
transport dm/dt which, according to Fick's law, is given by

$$\frac{dm}{dt} = v\rho = -D\frac{dn}{dz} \qquad (2.18)$$

where D is the diffusion coefficient of the solidifying species, ρ
its density and dn/dz is its concentration gradient in the plane
normal to the liquid solid interface. In the presence of the
solutal boundary layer dn/dt is modified (Nernst, 1904) to
$(n_o-n_i)/\delta$, where n_o is the bulk concentration, n_i the interface
concentration and δ is the width of the boundary layer. This leads
to an expression for growth rate v, valid for dilute solutions

$$v = \frac{D}{\rho} \cdot \frac{n_o-n_i}{\delta} \qquad (2.19)$$

    As in the case of vapour-phase boundary-layers, the simplest
approximation gives a linear growth rate inversely proportional to
the boundary layer width δ. In practical liquid-phase growth
systems, however, the crystal is usually rotated or a flowing
solution used, so that the boundary layer is reduced to maintain a
high growth rate.

    Carlson (1958) has shown that when the hydrodynamic
modification of the solutal boundary layer is taken into account,
the modified expression for the boundary layer thickness δ is

$$\delta = \left\{ 0.463 \left[\frac{\eta}{\rho_{sn}D}\right]^{1/3} \left[\frac{u\rho_{sn}}{\eta l}\right]^{1/2} \right\}^{-1} \tag{2.20}$$

where η is the solution viscosity and ρ_{sn} its density, l is the length of the crystal face and u the solution flow rate. Since $v \propto {}^1/\delta$ this expression indicates a relationship of the form

$$v = Au^{1/2} \tag{2.21}$$

Crystal rotation leads to a similar modification of the boundary-layer, whose width in this case is given by Burton, Prim and Schlichter (1953) as

$$\delta = 2^{2/3} D^{1/3} \eta^{1/6} \omega^{-1/2} \tag{2.22}$$

where ω is the angular rotation speed.

2.4 The Growth Surface

At high reactant flow rates or low substrate temperatures substrate/surface reaction processes limit the growth rate.

These include:

a) adsorption, dissociation and desorption at the growth surface

b) mobility of reactants and products on the surface

c) incorporation at lattice sites

All these processes are critically dependent upon the physical and chemical state of the growth surface and it is instructive to consider the way in which our knowledge of surface structures has expanded in recent years. Early theories of heterogeneous nucleation, eg that of Volmer (1939), considered the relative surface and volume free-energy changes associated with the formation of stable nuclei on flat surfaces. They showed poor agreement with

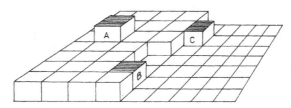

Fig.2.6   Possible lattice sites for the attachment of adsorbed atoms; A. Surface site; B. Ledge-kink site; C. Ledge site.

the widely variable experimental results on nucleation rate but
provided qualitative explanations of phenomena such as enhanced
nucleation on the walls of silica growth vessels, where macroscopic
surface roughening increased the nucleation probability. It was
soon realised that the microscopic condition of the growth surface
was frequently the dominant factor in determining nucleation rate.

2.4.1   The Kossel, Stranski, Volmer (KSV) theory

The importance of surface discontinuities as nucleation sites was
first recognised by Kossel, Stranski and Volmer (1939). Fig.2.6
shows possible attachment sites for adatoms at the growth surface.
The maximum binding energy between an adatom and the existing
crystal lattice occurs for incorporation at a kink in a surface
ledge, whilst at any point on the ledge the binding energy will be
greater than for an adatom attached to the flat surface. This
theory, together with those outlined below, form the basis of modern
crystal nucleation theory and have therefore been discussed in
detail in several texts (see, for example, Kaldis, 1974; Frank et
al, 1975). They are reviewed in outline here to provide a
background against which to discuss more recent developments and
experimental results.

2.4.2   The Burton, Cabrera, Frank (BCF) theory

One of the major drawbacks of early nucleation theories was
that once the kinked ledge had received sufficient adatoms to
move it to the edge of the crystal it could no longer function as a
low energy nucleation site. Generation of a new ledge then required
type A (Figure 2.6) adsorption, for which the calculated required
supersaturation was typically around 50%. In their celebrated
paper, Burton, Cabrera and Frank (1951) showed how the emergence
points of dislocations with screw components at crystal surfaces
acted as continuous generators of surface ledges (Figure 2.7)

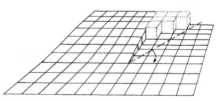

Fig.2.7   Schematic diagram of the growth surface showing nucleation
           sites (broken lines) at the surface step produced by an
           emergent screw dislocation. Continued nucleation causes
           the step to rotate in the arrowed direction, to the
           original position one atomic layer higher

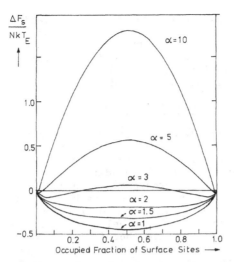

Fig. 2.8   Change in surface free energy as a function of surface
            coverage for various values of α (after Jackson 1958).
            Reprinted with permission from "Liquid Metals and
            Solidification", American Society for Metals, 1958, p181.

affording the possibility of continuous growth at much lower
saturations, as observed experimentally. The second important
concept discussed by Burton, Cabrera and Frank was that the ledge
and kink density, ie surface roughness, of a particular
crystallographic plane depended upon the interatomic bonding forces
between surface atoms and their neighbours.

Using Onsager's theories (Onsager, 1944; Wannier, 1945) of
cooperative lattice phenomena (and the Bethe (1935) many-body
approximation) which predict a smooth to rough transition as the
surface temperature is raised, Burton, Cabrera and Frank showed that
for an fcc lattice this transition temperature would be lower for
high-index planes than for low-index planes. For the latter the
theoretical transition temperatures obtained were larger than the
crystal melting point, suggesting that these planes remained
essentially smooth up to the melting point. It became possible to
understand, at least in principle, the experimentally observed
differences in growth rate on different crystallographic planes and
the formation of facetted crystals as a result of relatively slow
growth on the smooth low-index planes.

2.4.3 The Jackson α factor

One of the earliest attempts to quantify surface roughness was
made by Jackson (1958; 1975). If we consider a type A atom in

Figure 2.6 then the probability of it having a neighbour is proportional to the fraction of surface sites which are filled. The free energy change ΔF_s associated with filling a fraction x of the surface sites is given by

$$\frac{\Delta F}{NkT_E} = \alpha\, x(1-x) + x\, \ln x + (1-x)\ln(1-x) \qquad (2.23)$$

where N is the number of surface sites and $\alpha = L\xi/kT_E$

in which T_E is the equilibrium temperature at the liquid-solid transition and ξ is the fraction of the binding energy associated with the surface layer. ξ is always <1.0 due to binding of surface atoms to those in lower layers, but is largest for the closest-packed surface. In Figure 2.8 the free energy term $\Delta F_s/NkT_E$ is plotted against x for several values of α. From this we see that free-energy reduction only occurs for relatively smooth surfaces x<1% or x>99% for materials with large α-factors (eg organics or polymers), whilst for low α-factor materials, (eg metals) the maximum free-energy reduction occurs for x $\simeq$ 50%. Low α-factor materials are therefore less prone to facetting and show faster growth rates than high α-factor materials.

2.4.4 Periodic bond chain theory

The anisotropy factor introduced by Jackson in general terms was examined in crystallographic terms by Hartman and Perdok (1955) (see also Hartman, 1973) who classified crystal surfaces into F (flat), (stepped) and K (kinked) faces according to the number of periodic bond chains (PBCs) they contained. In this concept the

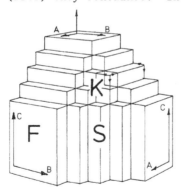

Fig.2.9   Hypothetical crystal with three periodic bond-chains: A ‖ [100]; B ‖ [010] and C ‖ [001]. Flat faces are (100), (010) and (001). Stepped faces are: (110), (101) and (011). Kinked face is (111) (after Hartman, 1973). Reprinted with permission from "Crystal Growth: an Introduction", Ed. P Hartman. North Holland, 1973.

atomic structure of the crystal is defined in terms of strongly-bonded
chains of atoms. Any crystal face containing more than two PBCs
(Figure 2.9) was a flat F face upon which nucleation was difficult and
growth therefore slow. S faces contained only one PBC. K faces contained no PBCs and were hence fast growing so that no 'face' developed.

Bennema and Gilmer (1973) reviewed the theories discussed above
in some detail and showed that they contain many parallel features.
They defined the change in Gibbs free energy ΔG resulting from the
roughening of a flat reference plane according to the relationship

$$\frac{\Delta G}{NkT} = \beta\left[\sum_{n=\infty}^{0} (1-c_n) - \sum_{n=1}^{\infty} c_n\right]$$

$$+ \alpha \sum_{n=-\infty}^{\infty} c_n(1-c_n)$$

$$+ \sum_{n=-\infty}^{\infty} (c_n-c_{n+1})\ln(c_n-c_{n+1}) \qquad (2.24)$$

where

$$\beta = \Delta\mu/kT \text{ and } \alpha = 4\epsilon/kT$$

(c_n is the concentration of sites in the nth atomic layer from
the flat reference plane which still have solid-solid bonds, $\Delta\mu$ is
the chemical potential difference between solid and fluid and ϵ is
the energy gain accompanying the replacement of a solid-solid bond
by a solid-fluid bond, ie roughening). Plotting log β against α
gives a curve dividing two growth fields of the form shown in Figure
2.10. In field A where the α value is high and the interatomic

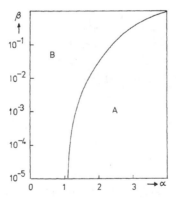

Fig.2.10   Curve dividing the region of positive values of the
           parameters β and α into sub-regions (A. stable; B.
           unstable) (after Bennema and Gilmer, 1973). Reprinted
           with permission from "Crystal Growth: an Introduction,"
           Ed. P Hartman. North Holland, 1973.

bonding forces are large, it is difficult to create rough surfaces
and the layer-growth mechanism predominates even for a relatively
high chemical or thermal driving force $\Delta\mu$. Where α is low, however
(or for moderate α values under conditions of high supersaturation)
the flat surface is unstable, is readily roughened and growth occurs
by a continuous random adatom absorption mechanism (field B). This
analysis, therefore, confirms the intuitive reasoning that the
roughening transition should be dependent upon both the driving
forces for the reaction (the chemical and thermal driving forces are
taken together as $\Delta\mu$) and the bond strength in the interface
region. The relative contributions are shown in Figure 2.10,
whereas earlier treatments lumped them together, predicting a fixed
value for the roughening transition.

Until this time much of the analysis of fluid-solid interfaces
had been based on the application of classical thermodynamics to
study energy and entropy changes associated with solidification.
The rapidly increasing availability of computing power in the late
1950s and early 1960s encouraged the development of microscopic,
stochastic models of the interface in which growth is expressed as
the changing occupancy of sites on a three-dimensional lattice
structure, the filled part of the lattice representing the solid
material. The starting point for many early analyses was the Ising
model of the crystal lattice originally applied to a magnetic system
(see, for example, Mattis, 1965 and the 'two-level' model
subsequently derived by Mutaftschiev (1965)). Notable advances in

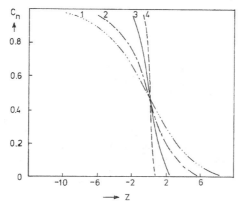

Fig.2.11. Concentration distribution of "solid" atoms, c_n producing
a minimum of the free energy of the system for $\beta = 0$ and
certain values of the parameter $\alpha = 4\xi/kT$ ((1) $\alpha = 0.446$,
(2) $\alpha = 0.769$, (3) $\alpha = 1.889$, (4) $\alpha = 3.310$) as a function
of position, z, across the growth interface (after Bennema
and Gilmer, 1973). Reprinted with permission from "Crystal
Growth: An Introduction," Ed. P Hartman, North Holland, 1973.

development of such theories were made by Temkin (1966, 1971) who
showed that as α decreased, the interface depth increased from one
or two atomic planes (smooth) to several (rough) (Figure 2.11).

2.4.5 The Müller-Krumbhaar model

One of the major limitations of early stochastic theories was
that their formulation did not take into account the nature of the
adsorption process and assumed that any difference in adsorption and
desorption rates was only reflected in a change in growth kinetics
rather than the actual incorporation process. Working from a
modified Ising model with the solid-above-solid restriction,
Müller-Krumbhaar (1974) defined the rate at which lattice sites S_j
would be filled according to the equation

$$\frac{d}{dT} <s_j> = -\left[(<s_j> - <s_{j+1}>\bar{w}_{j-}) + (<s_j> - <s_{j+1}>\bar{w}_{j+}) \right] \qquad (2.25)$$

$$\underbrace{\hspace{3cm}}_{\text{desorption}} \qquad \underbrace{\hspace{3cm}}_{\text{adsorption}}$$

where $\bar{w}_{j-}$ and $\bar{w}_{j+}$ are the conditional probabilities that the site
variable S_j changes its sign. $\bar{w}_{j\pm}$ is in turn given by

$$\bar{w}_{j\pm} = \frac{1}{2\tau} (1\pm\tanh E_j) \qquad (2.26)$$

where τ is a characteristic time constant and E_j the energy
associated with the transition is given by

$$\bar{E}_j = \frac{1}{kT_B} (4J<s>_j + \Delta\mu) \qquad (2.27)$$

Having expressed the adsorption-desorption transition
probabilities in this way, the effect of changing the ratio between
them can be studied. This is shown in the plot of surface roughness
$\bar{R}$ against growth rate $\bar{V}$ for a series of values of β, a parameter
derived from the absorption/desorption ratio (Figure 2.12). From
this we see that, whilst growth rate increases with increasing
roughness, there is no unique law defining this relationship and the
growth process is critically dependent upon the detail of the
mechanism of adsorption. To a first approximation, however, a power
law of the form

$$\bar{R}(\bar{V}) - \bar{R}(0) \simeq \bar{V}^2 \qquad (2.28)$$

can be derived, explaining both the general form of earlier workers
results and the wide variation in the power law exponent obtained.

A further important result of this analysis is the prediction
of a periodically varying growth velocity and surface roughness, the

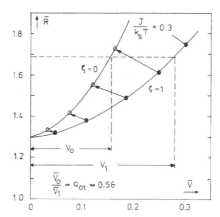

Fig.2.12 Average interface roughness $\bar{R}$ versus velocity $\bar{V}$ for two
values of ξ. The points connected by arrows correspond to
the same chemical potential difference (after Müller-
Krumbhaar, 1974). Reprinted with permission from Phys.
Rev.B published by the American Physical Society, New York.

amplitude of which decreases as the average velocity increases
(Figure 2.13).

So we see that the gradual evolution of nucleation and growth
theories leads from the simple postulations of rough and smooth
growth surfaces (with predicted growth rate $V \propto \Delta\mu$ in the rough case
and $V \propto \Delta\mu^n$ in the two-dimensional layer growth case) to a very
complicated model in which surface roughness and the adsorption-
desorption processes are inter-related. Whilst explaining the
variability of experimental observations on growth rate, this is
hardly helpful. Even the question of whether or not a roughening
transition occurs as the interface temperature rises is a source of
considerable debate (Müller-Krumbhaar, 1976). The problem is
further compounded when we consider that the simple theories described above treat the adsorption-desorption process as attachment or release of a single atom or molecule, whereas in practical
systems a multibody approach may well be more appropriate.

Later work by Gilmer's group (eg Gilmer and Broughton, 1983)
suggests that a definite roughening transition occurs within the
energy region

$$0.59 \; < \mathrm{kT}/\varphi < 0.66 \qquad\qquad (2.29)$$

where T is the surface temperature and φ is the bond energy. Such
transitions are revealed by Monte Carlo simulations and molecular
dynamic models, which do not impose the same constraints on atomic

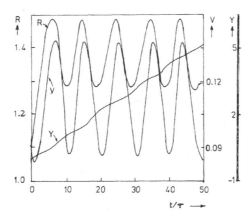

Fig.2.13   Surface roughness R, velocity V and position Y as a
           function of reduced time t/τ (after Müller-Krumbhaar,
           1974). Reprinted with permission from Phys.Rev.B pub-
           lished by the American Physical Society, New York.

configurations as those found in earlier models (eg Ising). The
important conclusion derived by Gilmer was that the near-perfect
surface can only be achieved in the presence of a very small
chemical driving force.

    So far, experimental evidence for defined transition
temperatures for surface roughening events are confined to the ^{4}He
configurations as those found system at temperatures <2°K. The
results are summarised by Mutaftschiev (1983) as follows:

                T > 1.5K   No singular faces (ie rough)

        1.15K > T > 0.85K   Basal face (0001) stable

        T < 0.85K   (1000)(1120) prismatic faces
                    + (0001) basal faces

        T < 0.365   Basal, prismatic and (1120)
                    pyramidal faces stable

    Although most early models assume a "hard-sphere" adatom, or
pair-wise interactions, Tiller (1984) has recently demonstrated the
need to use multibody effects to calculate the potential energies
associated with the structural and kinetic aspects of the growth
surface. His analysis is based on the definition of the nucleus in
terms of the potential energy Φ and atomic positions relative to a
coordinate origin v_{ij} as given by the Born-Oppenheimer expression

$$\Phi(x_1 \ldots x_n) = \frac{1}{2!} \sum_{i \neq j} \sum V^{(2)}(r_{ij})$$

$$+ \frac{1}{3!} \sum_{i \neq j \neq k} \sum \sum V^{(3)}(r_{ij}, r_{ik}, r_{jk}) + \ldots$$

$$+ \frac{1}{n!} \sum_{i \neq \ldots \neq n} \ldots \sum V^{(n)}(r_{ij} \ldots, r_{in} \ldots) + \ldots \qquad (2.30)$$

$V^{(2)}$ is the Mie potential (generalised Lennard-Jones potential) given by

$$V^{(2)}(r_{ij}) = \frac{\epsilon}{m-n} \left[n \left[\frac{Ro}{r_{ij}} \right]^{m-n} \left[\frac{Ro}{r_{ij}} \right]^{n} \right] \qquad (2.31)$$

$V^{(3)}$ are three-body Axilrod-Teller potentials given by

$$V^{(3)}(r_{ij} \ldots) = Z \left[\frac{1 + 3\cos\theta_{ij} \cos\theta_{ik} \cos\theta_{jk}}{(r_{ij} r_{ik} r_{jk})^{3}} \right] \qquad (2.32)$$

The incorporation of the three-body terms resolves many of the discrepancies between the simple two-body results and practical experience. Thus, whilst two-body calculations indicate that trimers will always appear as planar equilateral triangles and tetramers as tetragonally-bonded clusters, in practice linear molecules are frequently found, eg Si_3, C_3, C_4, SiC_2 etc. Again, in the modeling of surface relaxation effects, pair models frequently

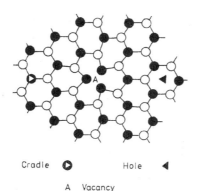

Cradle ◑   Hole ◀

A Vacancy

Fig.2.14   Relaxed equilibrium geometry of a surface vacancy. Also shown are the cradle and hole adatom adsorption sites (after Pearson et al, 1984). Reprinted with permission from J.Cryst.Growth published by North Holland, Amsterdam.

indicate an increase in interatomic spacing between first (surface) and second layers, whereas the reverse is generally true. When three-body terms are included in the potential energy function, the results (Halicioglu, 1980) agree closely with experimental observation. Halicioglu's results also showed that the depth of atoms involved in the relaxation increases as the degree of close-packing decreases, ie in the order (110), (100), (111).

A practical example of the use of three-body potential functions in the determination of surface reconstruction is given by the analysis of {111} Si and SiC surfaces by Pearson et al (1984). They showed that on cleaving the (111) Si face the compressive stress generated on reconstruction is partially relieved by the formation of surface ledges and surface vacancies, the latter frequently in the form of surface Frenkel-pairs. Further adatom sites are generated by the appearance of holes along the ledge face as reconstruction occurs. The calculation of the potential energy distribution confirms that cradle sites (Figure 2.14) and hole sites are favourable adsorption points on the surface. Cradle, hole and single vacancy formation energies are calculated as 0.73, -0.27 and -1.04 eV respectively, so that vacancy-hole pair formation (-1.31 eV) is energetically favourable, particularly if the individual defects are located on adjacent sites to give a surface Frenkel pair (-1.59 eV). A second observation is the non-equivalence of (111) and ($\bar{1}\bar{1}\bar{1}$) in the SiC zinc-blende lattice where the Si-rich face suffers a 26% contraction with a corresponding surface-free energy γ_S = 2220 erg/cm^2 compared with a 59% contraction of the C face (γ_S = 300 erg/cm^2). In consequence the orthogonal ($\bar{2}11$) SiC face is more stable than ($2\bar{1}\bar{1}$) SiC whereas in Si the reverse is true. Insights into the nature of the nucleation site and the compressive or tensile nature of the surface are important in explaining the adsorption of precursor, impurity and dopant atoms and in studying the surface defect structure which determines their mobility.

The importance of multi-body terms in nucleation theory is further reinforced by the practical and theoretical studies undertaken by Fink and Ehrlich (1981) and Fink (1982) to study the adsorption of Re atoms on a tungsten-surface. Pair interaction is found to be repulsive, whereas Re trimers in both linear and triangular configurations are stable. Tetramers and pentamers with 'open' structures were also observed and in some cases the surface mobilities of such clusters were found to be higher than those of single Re atoms on the (110) surface.

So we see that modern theories of surface atomic structure involve much more complex structures and more complex interactions with adatom species than those proposed in the early theories of Volmer and co-workers. They can be used to explain many commonly observed growth effects such as the non-equivalence of different crystallographic surfaces in terms of growth rate (eg van Erk et al, 1980; Morgan, 1974) and differing impurity incorporation rates in

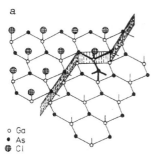 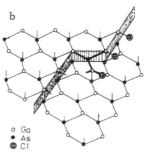

Fig.2.15   Chemisorption of a Ga-Cl-As complex in a kink on a) a
(111) Ga surface and b) a ($\overline{1}\overline{1}\overline{1}$) As surface. In the former
case adsorption does not impede ledge growth whereas on
the latter decomposition must occur before further ledge
growth is possible (after Hottier and Theeten, 1980).
Reprinted with permission from J.Cryst.Growth published
by North Holland, Amsterdam.

facetted and non-facetted regions of the growth surface (eg Mullin
and Hulme, 1960). In the latter case, radio-tracer experiments
established that Te dopant incorporation in InSb was increased by a
factor of six on {111} facet planes compared with the matrix value.
This result was attributed to the difficulty of nucleation on facet
planes leading to a higher supersaturation and fast lateral growth
rate. This in turn led to the entrapment of a higher concentration
of Te adatoms than in off-facet regions, where the equilibrium
segregation coefficient was observed.

    A further interesting example of the relative importance of
different nucleation sites comes from the work of Cadoret and
Cadoret. Figure 2.15 shows the differing orientations of a
Ga-Cl-As complex adsorbed onto kinks in (111) and ($\overline{1}\overline{1}\overline{1}$) surfaces.
On (111) surfaces the complex does not block the ledge growth
whereas in ($\overline{1}\overline{1}\overline{1}$) growth it must be decomposed before further growth
can occur.

2.5   Adsorption–Desorption

    In vapour phase growth the concentration of adsorbed atoms on
the growth surface is determined by the partial pressure of the
vapour and the surface temperature. According to the nature of the
adatom growth-surface interaction, the adatom may be strongly bonded
and immobile, giving rise to a Fowler-type adsorption isotherm
(Fowler, 1962) or weakly bonded to produce a mobile isotherm of the
Hill-de Boer type (de Boer, 1968). For good crystal perfection in
the growing layer, the adatom must be capable of migration to
suitable nucleation sites or of moving to new sites to eliminate
lattice defects. In other words, the mean diffusion length $\bar{x}$ must

be well in excess of the nucleation site spacing. If the surface
diffusion coefficient of the adsorbed species is D_s, ΔH_a its heat of
adsorption and ν_s its surface vibrational frequency.

$$\bar{x} = D_s \nu_s^{-1} \exp \ (-\Delta H_a/RT) \qquad (2.33)$$

For a given adsorbate and growth temperature T, $\bar{x}$ can be
increased either by raising the vibrational frequency of the adatom
or by increasing the surface diffusion coefficient. For this
reason, judicious choice of adatom species or the provision of
additional localised energy through plasma or photon enhancement is
often beneficial in surface-kinetically limited reactions.

In determining the optimum adsorbate structure for crystal
growth we therefore need information on both the concentration
(fractional surface coverage, θ) and geometrical shape of the
adsorbate nuclei. If the concentration is low then migration
distances are larger and growth rates slow, whereas very high
condensation rates do not allow time for atomic rearrangement of the
first deposited layer prior to overgrowth, so leading to high defect
concentrations. Similarly, if the adsorbate incompletely 'wets'
the growth surface 3D nuclei are produced where 2D nuclei are
required for layer growth. The shape of the nucleus is determined
by the sign of $\Delta\gamma$, the change in specific surface free-energy on
formation of the nucleus. This is given by the relationship

$$\Delta\gamma = \gamma + \gamma_{ns} - \gamma_s \qquad (2.34)$$

where γ is the specific surface free-energy of the nucleus, γ_{ns} is
the specific surface free-energy of the nucleus/substrate interface
and γ_s is the specific surface free-energy of the original
substrate surface.

When $\Delta\gamma$ is zero the nucleus 'wets' the substrate perfectly,
becoming two-dimensional and therefore ideally suited for monolayer
propagation. For positive values of $\Delta\gamma$ the wetting is imperfect and
3D nuclei are produced. Negative values of $\Delta\gamma$ indicate a strong
interaction between adsorbate and interface. Mutaftschiev (1986)
has shown that the three cases $\Delta\gamma = 0$, $\Delta\gamma$ positive and $\Delta\gamma$ negative
can be related to the strength of the adsorbate-substrate bonding
(A-S). If A-A is equal to A-S then $\Delta\gamma = 0$, giving the ideal
condition for crystal growth. If A-A > A-S, $\Delta\gamma$ is positive and no
nucleation takes place until the gas phase is saturated.
Supersaturation leads to preferential 3D nucleation on the growth
surface simply because the required critical volume is reduced in
relation to that required for nucleation in the gas phase. If A-A <
A-S, $\Delta\gamma$ is negative and 2D condensation can occur below the
saturation pressure of the gas phase, due to the surface free-energy
reduction associated with the chemical bonding between nucleus and
substrate.

This type of analysis promises to be of significant assistance in determining the correct vapour pressure required and adsorbate species in semiconductor epitaxial heterostructure growth, and also provides a physical interpretation of earlier thermodynamic theories.

For example, the relationship between vapour pressure and surface coverage has been shown by Chakraverty (1973) to be of the form

$$\frac{p}{p_0} = k_2 \frac{\theta}{1-\theta} e^{\theta/(1-\theta)} . e^{-k_1 \theta} \qquad (2.35)$$

In this expression, k_1 is independent of substrate, but dependent on the interaction strength between adatoms (A-A), whereas k_2 is a measure of the adsorption strength of the surface (A-S). The isotherms for $k_2 = 0.1$ to 10 ($k_1 = 7$) shown in Figure 2.15 indicate that for small θ, $p/p_0 \simeq k_2 \theta$ but that, as k_2 decreases (adsorption increasing), condensation of a quasi-liquid layer with $\theta \approx 1$ occurs.

Unfortunately, although the general principles of sorption are well established (see for example Dushman, 1962), there have been relatively few studies of practical situations pertinent to crystal growth of semiconductors. Most experimental data are derived from studies on the condensation of metal atoms on the tungsten electrode of the field-ion microscope, which can then be used to register the presence of an adsorbed layer and the onset of nucleation (Hudson and Sandejas, 1967, 1968; Nguyen et al, 1972). The importance of the relationship between k_1 and k_2 and, in particular, the part played by k_2 in determining the nature of growth has been demonstrated by Dismukes and Curtis, 1973. They showed that when the adatom/substrate interaction is weak, as for example in the case of Au or Ag deposition on ionics such as NaCl or KCl, decahedral or icosahedral particle morphologies were observed, whereas strong adatom/substrate bonding (eg Ag on AgCl) leads to an f.c.c. cluster morphology and to good epitaxial growth.

In low-temperature deposition the reaction by-product desorption can frequently govern the reaction rate. In such cases the choice of carrier gas can significantly alter the reaction rate. Duchemin et al (1978) noted that in silane decomposition according to the reaction

$$SiH_4 \rightarrow Si + 2H_2 \qquad (2.36)$$

where

$$\frac{d(Si)}{dt} = k \frac{[SiH_4]}{[H_2]^{1/2}} \qquad (2.37)$$

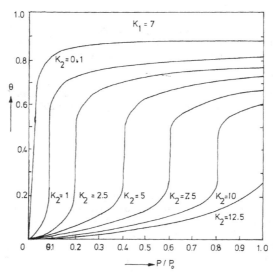

Fig.2.16   Isotherms for k_1 = 7 and k_2 = 1 to 10 (after de Boer, 1968). Reprinted with permission from "The Dynamical Character of Adsorption" by J H de Boer, published by Oxford University Press, Oxford, 1953.

the desorption of the H_2 product is retarded in the presence of an H_2 carrier gas and is inversely proportional to the H_2 carrier gas concentration. Wilkes, 1984, has observed a similar effect for the desorption of N_2 in the presence of an N_2 carrier gas during the reaction

$$SiH_4 + 4NO \rightarrow SiO_2 + 2H_2O + 2N_2 \qquad (2.38)$$

In an elegant treatment of the reaction kinetics of heterogeneous reactions he showed that the "effective activation energy" derived from the linear ln (growth rate) versus $1/T$ relationships was, in the case of strong product adsorption, a combination of the true activation energy ΔE for the reaction and a component due to the heat of sorption Q_s of the product species, such that the reaction rate dx/dt is given by

$$\frac{dx}{dt} = A\ p_0(1-Q_s)/RT\ \exp(-\Delta E/RT) \qquad (2.39)$$

These surface-kinetic limitations to growth rate and growth morphology become increasingly important as growth temperatures and pressures are reduced in an effort to produce thin multilayer structures with minimum interdiffusion and we shall return again to the subject in Chapter 5, in which we examine the role of surface reconstruction, stoichiometry, impurity adsorption and surface damage in controlling sorption, nucleation and growth.

CHAPTER 3

DEFECTS IN CRYSTALS

3.1 Introduction

Although the theoretical treatments of nucleation and crystal
growth make the simplifying assumption that the crystal lattice
beneath the growth surface is perfect, this is never the case. The
elevated temperature results in the formation of lattice point-
defects. As the temperature of the solid increases, so does the
amplitude of oscillation of the atoms on the lattice sites. An
increasing fraction of these are displaced from their sites to
produce lattice vacancies and interstitials. In the simple Frenkel
defect (Figure 3.1a) both are retained in the lattice, whereas in
the Schottky defect (Figure 3.1b) ejected atoms migrate to a free
surface or interface.

The concentration N of such lattice defects follows an
Arrhenius relationship of the form

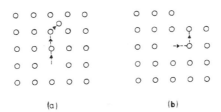

(a)   (b)

Fig. 3.1   a) A Frenkel Pair (interstitial + vacancy)
             b) A Schottky defect (the atom migrates to a free
                surface)

29

$$N = Ae^{(E_f/kT)} \qquad\qquad (3.1)$$

where T is the temperature of the solid and E_f is the formation
energy of the defect. In alloy semiconductors, point defects can be
generated in both cation and anion sub-lattices and the equilibrium
concentration will depend on both the lattice temperature and the
partial pressures of the alloy species over the solid. For example,
in GaAs grown at low p_{As} pressures the concentration of vacant
As-lattice sites greatly exceeds that of vacant Ga sites. For
growth under Ga-rich conditions the converse is true. The compound,
therefore, exhibits an extended range of composition (Figure 3.2)
from 49.91% Ga to 50.08% Ga at 1132°C (Hurle, 1979). In
$Cd_{0.2}Hg_{0.8}Te$ the high volatility of the Hg component leads to high
P_{Hg} (approximately 30atm at the melting-point 790°C) and an
extensive Te-rich non-stoichiometric range (Figure 3.3). The
defect capacity of each sublattice is a function of <u>local</u>
temperature, pressure and bond-strength; lattice breakdown may occur
well before the excess concentrations detectable by the X-ray
analysis techniques conventionally used to determine the phase
limit. In fact, <u>any</u> departure from stoichiometry may lead to atomic
clustering. These clusters will increase in size and stability as
the excess component concentration increases. There is, therefore,
a steady progression from isolated interstitials or vacancies to
coherent clusters (Guinier–Preston zones) and finally to incoher-

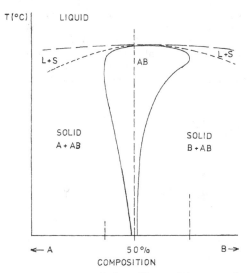

Fig. 3.2   Temperature-composition Phase Diagram for the A-B
binary system showing the extended phase-range of the
compound AB

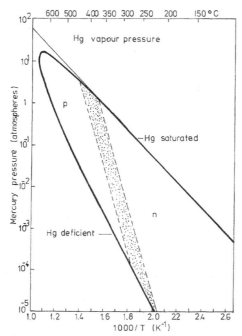

Fig. 3.3   Pressure-temperature Phase Diagram, showing the $Cd_xHg_{1-x}Te$
            solid phase-field.   (After Jones et al., 1980).   Reprinted
            with permission from J.Appl.Phys. published by American
            Institute of Physics.

ent precipitates. A further observation limiting the value of
the concept of a macroscopically-determined "phase-boundary" is
that the crystallinity of the material is also determined by the
extent to which the non-stoichiometry can be reduced by removal
of precipitating atoms.   This will depend upon the local temperature
and composition gradients which control migration rate and the

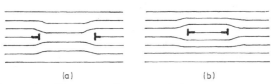

Fig. 3.4   Schematic diagram of crystal lattice planes showing:
            a) an intrinsic dislocation loop formed by vacancy
            agglomeration; b) an extrinsic dislocation loop resulting
            from interstitial agglomeration

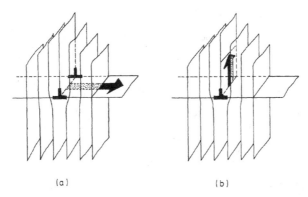

(a)   (b)

Fig. 3.5   Schematic diagram of an edge dislocation showing
            movement by a) glide and b) climb

proximity of surface or interfacial "sinks". Thus a thin epitaxial
layer may be satisfactorily grown under conditions of temperature
and pressure which would give rise to second-phase precipitation in
the interior of a bulk-crystal.

When the point defect concentration of the material is high,
cooling from high temperature produces a supersaturation. If the
cooling rate is high the excess is quenched-in but slower rates may
cause the individual defects to coalesce. Interstitials can
coalesce to form clusters and precipitates in the manner described
above, or in the case of alloy semiconductors, may be absorbed on
substitutional sites of the opposing sub-lattice to form "anti-site"
defects, eg Ga_{As}, a Ga atom on an As site. (It follows that an
increasing concentration of Ga_{As} antisites will lead to Ga
clustering on the As sub-lattice). Although vacancy clustering can
lead to spherical void formation, the surface free-energy is large
and the more usual situation is for the lattice to collapse locally
in the manner shown in Figure 3.4a to generate a dislocation loop.
Similarly the strain generated by the presence of an interstitial
cluster can be relieved by generation of such a loop (Figure 3.4b).
Dislocations may also be generated by applied thermal or mechanical
stresses to which the material is subjected during cooling from the
growth temperature. The properties and interactions of dislocations
have been reviewed in many books (eg Smallman, 1962; Hirsch et al,
1965).

Dislocations move through the crystal lattice by glide, moving
along a particular crystal plane known as the slip plane (Figure
3.5a), or by climb, the absorption or desorption of atoms from the

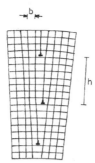

Fig. 3.6   Schematic diagram of a sub-grain boundary showing a
            periodic array of dislocations of Burgers vector b and
            inter-dislocation spacing h

constituent half-plane (Figure 3.5b). At high dislocation
concentrations the free-energy of the total array can be reduced by
"polygonisation" of the dislocations to form a planar array (Figure
3.6). The misorientation θ between the crystal lattices on either
side of the boundary is given by

$$\theta = \frac{b}{h}$$ \hfill (3.2)

where h is the inter-dislocation spacing and b is the Burgers
(displacement) vector of the dislocation. In addition to such
"tilt" boundaries, it is also possible to produce similar boundaries
by rotation of adjacent lattices (twist boundaries). Angles of
misorientation range from a few arc seconds to several degrees and
in view of the analogy with large-angle grain boundaries   such
polygonised dislocation arrays are commonly called "sub-grain"
boundaries. Despite the similarity in terminology, the origins of
the two defects are frequently quite different. Large-angle grain
boundaries can be nucleated by applied mechanical and thermal stress
during the process of solid-state recrystallisation and in this case
develop from subgrain boundaries as the misorientation increases
beyond the level which can be accommodated by primary dislocation
arrays. In normal growth of all but the softest semiconductor
materials, however, the main source is the breakdown of the growth
interface due to nucleation of solid of different crystallographic
orientations at different points on the surface or, in the case of
Bridgman growth (see Chapter 4) on the ampoule wall ahead of the
advancing primary growth surface.

    In this chapter we shall review first the origin and nature of
the macroscopic crystal defects arising from growth interface
breakdown. In the treatment of point and "extended" (line, planar

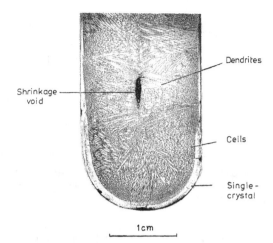

Fig. 3.7   Etched longitudinal cross-section through a quenched
HgTe ingot showing initial single-crystal growth at edge
followed by interface breakdown to give cells and
dendrites

or three-dimensional) defects the emphasis will be on the
interactions occurring between them, particularly in AB-type alloy
semiconductors where the need to preserve the structural integrity
of the A-B bonding leads to the introduction of "double-plane"
defects, ie the dislocation or vacancy loop "half-plane" must
accommodate both type A and type B atoms.

3.2   Growth Interface Breakdown

3.2.1   Cells and dendrites

In the preceding chapter we noted that defect-free growth could
only be achieved within a limited range of supersaturation. If the
rate of condensation of adsorbed atoms at the growth surface exceeds
the rate of incorporation on correct lattice sites, the growth
surface breaks down in the manner first described by Rutter and
Chalmers, (1953) to produce a cellular or dendritic structure. The
HgTe ingot, shown in Figure 3.7 was solidified from a Te-rich melt.
Solidification began at rapid rate at the ampoule wall and
single-crystal growth was maintained for the first 1-2mm of growth,
after which the accumulation of excess tellurium ahead of the
interface led to rejection into cell boundaries and finally to the
development of a fully dendritic structure. A central void
attributable to shrinkage on solidification is also visible. The

Fig. 3.8   Cross-section through a $Cd_{0.2}Hg_{0.8}Te$ ingot showing the columnar grain structure resulting from radial heat loss.

relationship between heat-flow and resulting dendritic structure has been modelled by Jones et al (1983) who showed that the dendritic structure was dependent upon the temperature distribution, cooling rate and the thermal conductivity of the liquid, solid and gas components of the casting system. The mechanisms of solute or alloy component rejection to cell and dendrite boundaries and the subsequent development of these structures is well understood and has been the subject of several reviews (see e.g. Rutter and Chalmers, 1953; Bolling and Tiller, 1961).

Unless a steep axial gradient is imposed, nucleation of solid on cooling occurs simultaneously at several points in the ampoule wall and the advancing solidification front moves inwards normal to the radial temperature gradient. The result is a "columnar" grain structure of the type shown in Figure 3.8. The grain boundary delineates areas within which dendrites share a common axis. When the solidification rate is reduced, the liquid-solid interface is planar rather than dendritic and each grain is then a true "single crystal". The art of single crystal growth is therefore to provide a single nucleation site, in the form of a "seed" or "substrate" crystal and to ensure that there is a sufficiently steep and uniform temperature gradient normal to the growth surface to prevent spurious nucleation at other locations. Even when the solid-ification rate is too slow, or the supersaturation too low, to generate cells and dendrites, single crystal nucleation begins simultaneously at several unrelated points on the ingot wall in the system shown in Figure 3.7, leading to inward growth of differently oriented single-crystals to produce a characteristic columnar crystal structure shown in Figure 3.8. Cooling in the absence of temperature gradients or under stirred conditions gives rise to an equiaxed grain structure whose average grain diameter increases as the cooling rate reduces.

3.2.2 Grain boundaries

In early theories, the atomically perturbed region of the boundary between single-crystal grains was held to be an amorphous region of atomic disorder, but more recently it has been shown by Bollman (1970), Chalmers and Gleiter (1971), Hu (1982) and others that the boundary can be described in terms of a "coincidence site lattice", CSL (Kronberg and Wilson, 1949), in which the extrapolation of the lattice sites of grain A onto those of grain B defines a new lattice of coincidence points. The periodicity $\sum$ of such points is defined as a multiple of the original lattice spacing.Figure 3.9 illustrates the case of a $\sum$ = 5 boundary with coincident lattice sites occurring on every sixth original lattice site. Alternatively, $\sum$ may be defined as the reciprocal of the fraction of total lattice sites occupied by coincidence sites. The degree of misfit increases with increasing $\sum$, an important consideration when determining the ability of the boundary to accommodate foreign atoms or act as a nucleation site for precipitation. Since boundary *energy* also increases with $\sum$ it is possible to reduce $\sum$ by the introduction of localised regions of greater disorder (grain boundary dislocations and/or boundary steps)

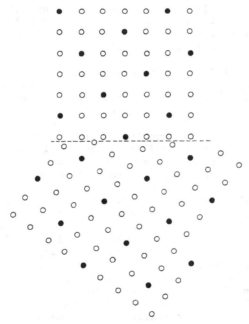

Fig. 3.9   Coincidence site lattice (CSL) with Σ=5 produced by overlap of two (100) lattices at an angle of 38°.

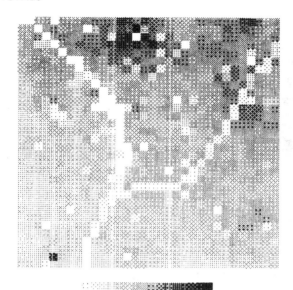

Fig. 3.10 A two-dimensional (32x32) infrared radiation detector
array showing the "fixed-pattern" noise due to variations
in zero bias resistance R_D between different detectors.
Minimum $R_0 = 2.9 \times 10^3$ Ω, Max $R_0 = 1.06 \times 10^5$ Ω, Mean $R_0 =$
4.45×10^4 Ω (courtesy of R A Ballingall and D J Lees, RSRE)

in order to produce a better coincidence ratio over the remainder of
the boundary. Whilst the CSL defines the degree of 'fitting' in the
regions between the dislocations or steps, the Burgers vectors of
the line defects themselves form a periodic array, frequently
referred to as the DSC lattice (Displacement Shift Complete). The
CSL/DSC grain boundary theory has been used successfully to explain
the nature and properties of many types of grain boundary (and also
inter-phase boundaries between lattices of different periodicity).
Good reviews of the subject are given by Warrington (1980) and
Balluffi et al (1982). Much of this work arose from studies of
grain boundary mobility and grain boundary precipitation during
recrystallisation and annealing of metals (eg Lorimer, 1975).
Unfortunately, little attempt has yet been made to apply it to the
study of grain boundaries in semiconductor materials. Such studies
as have been made on semiconductor materials indicate that grain
boundaries are generally detrimental, acting as charge carrier
recombination centres (Seager, 1982), charge leakage paths and sites
for enhanced impurity incorporation (Russell et al, 1981) and

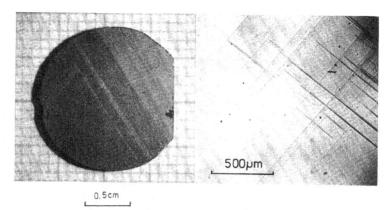

Fig. 3.11   Examples of twinning on: a) {111} planes in CdTe
b) {110} planes in $Ba_2NaNb_5O_{15}$

precipitation (Queisser, 1983).   Figure 3.10 shows the fixed-pattern
noise in a 32x32 element infra-red detector array induced by the
presence of a grain boundary and by individual dislocations and
precipitates in the $Cd_xHg_{1-x}Te$ material.

Although frequently regarded as a separate class of defect, the
twin boundary is a specialised form of grain boundary for which $\sum$ =
3.   Twins are occasionally observed in the predominantly covalent
semiconductors (eg Si, GaAs, InP, InSb) but are particularly
prevalent in materials with a high degree of ionicity - as, for
example, in CdTe (Vere et al, 1983) or complex oxides such as
$Ba_2NaNb_5O_{15}$ (Figure 3.11).   In such materials, small deviations from
the equilibrium bond configuration lead to rapid rise in energy and
the preservation of lattice coherence assumes greater importance.
Adsorbed atoms on the growth interface are therefore incorporated on
one of two equivalent configurations eg (111) or ($\bar{1}\bar{1}\bar{1}$).   If "island"
growth occurs simultaneously from several nucleation sites and
adjacent islands comprise (111) and ($\bar{1}\bar{1}\bar{1}$) orientations respectively,
then each coalescence boundary will be a 180° rotational twin.   The
conditions which favour island growth (high impurity concentration
or rapidly fluctuating growth rate) will therefore lead to enhanced
twin-boundary formation.

3.2.3   Growth striations

Small quantities of impurity, or excess or deficiency of native
species in compound semiconductors, can be accommodated in the
crystal lattice without breakdown of the growth surface.   However,
there is a high probability that the incorporation will be spatially

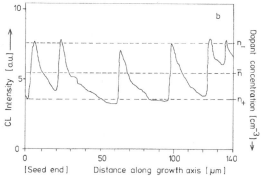

Fig. 3.12   a) Growth striae perpendicular to the growth axis of a
CZ-grown, Ge-doped InP crystal revealed by cathodoluminescence. b) Corresponding line-trace of cathodoluminescence intensity, which is inversely related to
dopant concentration (after Warwick et al, 1983). Reprinted with permission from J.Cryst.Growth published by
North Holland, Amsterdam.

there is a high probability that the incorporation will be spatially
non-uniform. This occurs because local fluctuations in temperature
at the growth surface due to thermal or solutal convection (see
Chapter 4, Section 4.3) produce oscillations in the growth rate.
These, in turn, lead to variations in stoichiometry and impurity
uptake. Alternatively such striations may arise through crystal
rotation in an asymmetric thermal field. Growth striations and
their origin have been the subject of many studies (see, for
example, Landau, 1958; Schlichter and Burton, 1958; Carruthers,

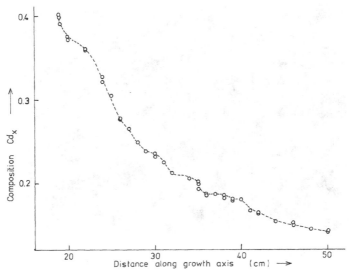

Fig. 3.13 Axial composition profile of Bridgman-grown $Cd_xHg_{1-x}Te$ showing periodic fluctuations superimposed on the 'normal-freeze' profile

1967; Hurle, 1967; Cockayne and Gates, 1967; Witt and Gatos, 1967; de Koch, 1976).

The sharpness of the striations will depend upon the extent of 'melt-back' of the growing interface on encountering hotter liquid, or on the post-growth interdiffusion in the solid state and hence upon growth rate, temperature gradient and the diffusion coefficient for the particular material. Figure 3.12a shows such striae in a Czochralski-grown, Ge-doped ingot of InP (Warwick et al, 1983). The associated concentration profile, determined from cathodoluminescence studies, is shown in Figure 3.12b. For a mean Ge-doping level of $1.0 \times 10^{19} cm^{-3}$, the doping concentration across a single striation ranges from $8.9 \times 10^{18} cm^{-3}$ to $1.12 \times 10^{19} cm^{-3}$. The steep leading edge of the profile is due in this case to interface melt-back. The striation spacing is about $25-30 \mu m$. For comparison, Fig 3.13 shows the variations in Cd concentration down a Bridgman-grown ingot of $Cd_xHg_{1-x}Te$. In this materials system the large segregation coefficient $(k \approx 3)$ leads to a monotonic decrease in Cd concentration as growth proceeds. Superimposed on this are short-range fluctuations imposed by periodic variations in boundary layer composition. In this case, however, the high diffusion coefficient combined with slow growth rate ($\sim 1mm\ hr^{-1}$) and shallow temperature gradients across the solid–liquid interface produce a much larger striation periodicity of 2-3 mm.

3.3 Lattice Point-defects

Much of the information on isolated point-defect species
(impurity atoms, and native vacancy and interstitial defects) is
derived from measurements of their related electronic transport
properties, particularly as revealed by the Hall effect. Unfortunately the technique records only the net charge-carrier
concentration $|N_D-N_A|$, where N_D is the donor and N_A the acceptor
concentration respectively, rather than the actual concentrations of
lattice defects producing them. Whilst some assumptions on the
ionisation state of the defect can be drawn from the rate of
'freeze-out' of the charge carriers as the temperature decreases, in
many cases complicating factors such as the spatial non-uniformity
and variation in charge states, the presence of residual background
impurities and surface or interface charge states lead to
uncertainties in the results. Nevertheless, some general trends are
expected. Deviation from stoichiometry toward the cation-rich side
in III-V or II-VI compounds results in an excess donor concentration, whilst excess of the Group V or Group VI element should lead
to hole generation and p-type material (Figure 3.14). In the presence of residual donor impurities the p-n transition is shifted

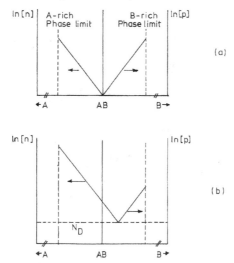

Fig. 3.14 Variation in carrier concentration with deviation from
           stoichiometry a) pure material; b) material containing
           residual donor impurity

to the anion-rich side of stoichiometry, whilst acceptor impurities shift the transition in the cation-rich direction; in practice the idealised situation shown in Figure 3.14a is rarely observed.

In II-VI materials, determination of the true point-defect concentration is further complicated by the possibility of self-compensation effects, in which the introduction of a donor impurity leads to the *automatic* generation of a compensating native acceptor species (Kroger and Vink, 1956; Mandel, 1964; Marfaing, 1981). Whether or not self-compensation occurs is dependent upon the net energy change resulting from two conflicting processes: the energy required to produce the compensating native defect and the reduction in energy resulting from the recombination of a free carrier at the new defect. The latter is related to the band-gap of the material, whilst the native defect formation energy is related to the bond-strength or cohesive energy of the material. Self-compensation is therefore more likely in materials where the band-gap/cohesive-energy ratio is large, ie II-VI and ionic materials. It is less likely in III-V materials. In fact, the evidence for the frequent and dominant occurrence of self-compensation has been weakened by the findings of recent research, in which many of the energy levels formerly attributed to native acceptor species have now been identified as relating to residual acceptor impurities (Dean et al, 1978; Marfaing, 1981) and the observed discrepancy between free carrier concentrations measured at high temperature and those observed after quenching to room temperature has been satisfactorily explained by co-precipitation of excess anion or cation and impurity species (Magee et al, 1975; Selim et al, 1975; Bensahel et al, 1979; Pautrat et al, 1982). Furthermore, in the few cases where native-defect impurity complexes have been observed, eg V_{Zn}-Cl (Schneider, 1967) the energy level is much deeper than that previously attributed to the native acceptor.

The interaction between point-defects and dopant atoms in GaAs has been studied in depth by Hurle, 1978, 1980; Newman, 1984, 1985; van Vechten, 1984 and others. The reactions involved are complex and highly sensitive to temperature ranges, cooling conditions and the stoichiometry or growth mode of the starting material. Using local-mode infrared spectroscopy, Maguire et al, 1985, have shown that site-switching of boron or silicon dopant atoms occurs on annealing GaAs. In the case of silicon, Si_{Ga} defects are converted to Si_{As} defects on annealing at 350°C in a reaction involving As vacancies

$$Si_{Ga} + V_{As} \rightarrow Si_{As} + V_{Ga} \qquad (3.3)$$

Annealing at higher temperatures promotes the reverse reaction. In the case of the amphoteric Group IV dopants, Ge and Sn, Hurle (1978), found excellent correlation between the experimentally determined carrier concentrations and his theoretical model based on the assumption that the compensation mechanism involves Ge donors on the Ga sub-lattice (Ge^{+}_{Ga}), Ge acceptors on the As sub-lattice (Ge^{-}_{As}) and acceptor-like complexes formed between the donor and a gallium vacancy ($Ge_{Ga} V^{-}_{Ga}$). More complex reactions involving the generation of "split interstitial" defects of the form V_{Ga}-Ga_i-Te_i (Dobson et al, 1979) and Ga_{As} "anti-site" defects (van Vechten, 1975) have been invoked to explain the electrical and lattice super-dilation effects observed on doping with Te. Although no definitive lattice-site data are available yet for II-VI compounds, Capper et al (1982, 1985) have found that whilst some $Cd_xHg_{1-x}Te$ dopants behave in a predictable manner, giving carrier concentrations proportional to their concentration, this is frequently not the case for atoms which can occupy more than one lattice site (interstitial or substitutional on either sub-lattice). In such cases the relative concentrations of native defects on each sub-lattice appear to play a significant role in determining the observed electrical properties.

So we see that the simple picture of individual, isolated impurity atoms, native vacancy or interstitial sites is a gross over-simplification of the real situation. In reality the interactions between these classes of defects can lead to complex combinations; the exact constitution and electronic charge state depend upon the crystal growth and annealing conditions, the defect species available and the local lattice environment. For this reason 'averaged' values of transport and luminescence properties are of limited value in describing defect complexes. In modern research, corroborative evidence is sought from the parallel use of techniques such as local mode IR spectroscopy and Optically Detected Magnetic Resonance (ODMR) which provide data on the symmetry of the defect complex. Cathodoluminescence is also increasingly used to provide information on the spatial distribution of emission and recombination centres.

3.4 Dislocation Lines, Loops and Precipitation

The subject of dislocations in semiconductor materials has been thoroughly covered in reviews and books (see, for example, Ralph et al, 1980; Baglin et al, 1984; Cullis and Joy, 1981; Cullis et al, 1983; Cullis and Holt, 1983). Here we consider only the general ways in which they arise and the characteristic distribution in bulk ingots and epilayers. Earlier works tended to treat second phase precipitation as a separate class of defect, but much recent work suggests that there is a definite link between precipitate and dislocation structures and they are therefore considered together here. We also emphasise the special relevance to modern crystal growth theory, of dislocations generated at the interface between substrates and epitaxial layers, either through lattice mismatch or incorrect substrate surface preparation.

1 cm

Fig. 3.15   Cross-section of a CZ-grown silicon crystal showing the
            "swirl pattern" produced by the precipitation of oxygen
            in the form of coesite, SiO_2 (courtesy of K Barraclough,
            RSRE)

If a crystalline material is cooled from high temperature, the
equilibrium point defect concentration characteristic of that
temperature must be reduced to that of the end point temperature.
Provided the cooling rate is slow enough, the point defects can
migrate to dislocation, grain boundary or surface sinks. If a high
rate of cooling is used, the individual point defects are quenched
or frozen into the lattice, giving a high elastic strain. Cooling at
intermediate rates or annealing of fast-cooled material causes
agglomeration of the defects to form small voids (in the case of
vacancy condensation) or clusters (in the case of interstitial
condensation). The free energy of the configuration can be reduced
(Kuhlmann-Wilsdorf and Wilsdorf, 1960) by collapse of these defects
to form planar vacancy or interstitial defects bounded by a
dislocation loop (Figure 3.4). Initially these loops may be only a
few Angstroms in diameter but, once formed, act as a sink for
further condensation and in some materials can grow to several
hundred microns in diameter.

In semiconductor materials such point defect clusters were
first observed when improvements in crystal growth techniques in the
late 1950s led to the development of dislocation-free germanium
(Tweet, 1959) and silicon (Plaskett, 1965). Using copper and
lithium decoration techniques (Dash, 1956; Takabatake et al, 1970)
combined with Sirtl etching (Sirtl and Adler, 1961) and X-ray

topography, de Koch (1973) showed that the microclusters were
distributed in a pattern corresponding to growth striations. Slices
cut normal to the growth direction, therefore, showed characteristic
'swirl' patterns (Figure 3.15). de Koch identified two types of
defect, the larger of which he called Type A and the smaller Type
B. The average concentration of the large A clusters was 5×10^5 to
$2 \times 10^6 cm^{-3}$ and appeared independent of growth atmosphere (argon or
vacuum). Type A clusters were not observed in the surface region (a
"skin" depth of approximately 2mm). Type B concentrations in this
region were about $2 \times 10^6 cm^{-3}$ compared with 7.5×10^5 - $3 \times 10^7 cm^{-3}$ in the
bulk. The nucleation of both types of cluster was attributed to
precipitation of vacancies at heterogeneous lattice sites provided
by vacancy-oxygen complexes. However, later work (Patel et al,
1977) suggests that the microdefects are intrinsic faults composed
of Si interstitials generated during O_2 precipitation, since for
every two oxygen atoms incorporated, one Si interstitial must be
formed if excessive lattice strain is to be avoided. The situation
is complicated by the variation in oxygen concentration with
position in the ingot (Barraclough, 1982) and dopant concentration
(Barraclough, 1985) and by the coexistence of oxygen and carbon as
residual impurities (Kolbesen, 1982). Further complications are
introduced by the effects of differing thermal treatments on the
density of the pre-existing nuclei and the generation of faulted
dislocation loops around the precipitate. The extrinsic fault
(Ponce et al, 1983) is necessary to accommodate the excess
interstitials generated by the oxygen precipitation. The morphology
of the precipitate is dependent upon the anneal temperature, being
rod-like (Bourret, 1984) for temperatures in the range 450°C-650°C,
{100} platelets (Ponce et al, 1983) for temperatures in the range

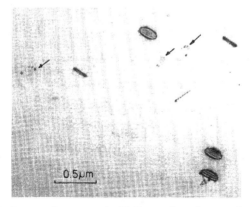

Fig. 3.16   Precipitate-decorated dislocation loops in as-grown
            GaAs. Arrows indicate "out-of-contrast" loops which show
            the effect most clearly (after Stirland et al, 1978).
            Reprinted with permission from J.Mater.Sci. published
            by Chapman and Hall, London.

650°-950°C and polyhedral precipitates of amorphous silica at
temperatures above 950°C. A summary of the defect structures
resulting from various conditions has been given by Claeys et al
(1982) and is covered extensively by Huff et al (1981) and Goorissen
(1982), and in a more recent review article by Ponce (1985). Despite
the incomplete understanding of the mechanisms involved, the defect
studies have been used to good effect in the technologically
important "intrinsic gettering" process. Rozgonyi et al (1976)
showed that by careful choice of annealing conditions it is possible
to produce a defect-free 'denuded' zone in the surface region of the
slice whilst using the stacking faults in deeper regions to "getter"
transition metal impurities from the zone.

      In alloy semiconductors loop formation is complicated by the
fact that the dislocation half-plane contains both anion and cation
species. Loop formation in III-V or II-VI materials is usually
accompanied by precipitation of the group V or group VI element.
The formation of such defects in GaAs has been discussed by
Stirland, et al (1978) and in $Cd_xHg_{1-x}Te$ by Schaake and Tregilgas
(1983) and Williams and Vere (1986). Stirland et al were able to
analyse some of the larger loops in Czochralski-grown GaAs. The
700-2000Å diameter loops (Figure 3.16) were found to be extrinsic in
character, bounded by an edge dislocation, which was decorated at
intervals around the periphery by small precipitate particles. The
nature of the precipitate found in these studies was unclear. Since
the specimens were heavily doped with Te or Se, $GaTe_3$ or $GaSe_3$
precipitates were possible. Similar precipitation had been observed
on loops on Sn doped material by Titchmarsh and Booker (unpublished
work) and tentatively ascribed to Sn precipitation. In more recent
studies of undoped gallium arsenide (Augustus and Stirland, 1980;
Cullis et al, 1980; Lodge et al, 1985; Stirland, 1986) As
precipitation on dislocations was detected.   In CdTe and
$Cd_xHg_{1-x}Te$, where undoped material has been studied, evidence of
excess Te precipitation has been found on dislocation lines and
loops (Schaake et al, 1985; Williams and Vere, 1986), and the
surrounding area is denuded of Te precipitation.

      In growth techniques involving confinement of the solidified
material in a crucible or similar container, expansion of the solid
against the rigid wall or uneven contraction (as, for example, when
areas of the solid surface adhere to the container wall) also act as
sources of elastic strain. Once the strain exceeds the elastic
limit, dislocations will be generated in these regions and propagate
into the bulk of the crystal. In Czochralski growth, although the
material is not under stress from a container, the high thermal
gradient in the solidifying ingot (which will be the subject of
further discussion in Chapter 4) acts as source of elastic strain,
which is again relieved by dislocation generation. Dislocations
produced by radial stresses usually propagate primarily by glide
along the {111} planes of the cubic lattice to give characteristic

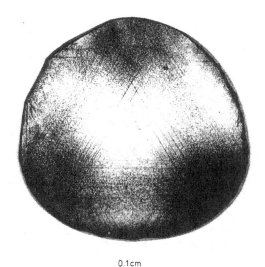

0.1cm

Fig. 3.17   Cross-section through an LEC-grown InP ingot showing the
            triangular slip pattern produced by dislocation glide on
            inclined {111} planes (courtesy of J B Mullin and B W
            Straughan, RSRE).

star patterns (Figure 3.17). A W-shaped dislocation density
distribution as a function of crystal diameter is characteristic of
this mechanism (Chen and Holmes, 1983; Blunt et al, 1982).

    A further source of dislocation generation in cooling crystals
is the presence of inclusions or second-phase precipitates. Such
particles act as stress raisers in the elastically-strained matrix
and give rise to dislocation generation by prismatic punching

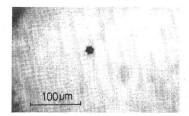

Fig. 3.18   Prismatic punching of dislocations on the {111}<110> slip
            system in CdTe to form a star-pattern around a central
            carbon-rich inclusion (a) transmission infrared
            micrograph; b) corresponding etch-pit pattern

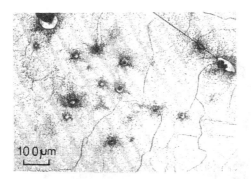

Fig. 3.19   Dislocation generation around second-phase tellurium
            inclusions in CdTe

(Hull, 1965). The threshold stress level at which this mechanism
operates is dependent on particle size and compressibility, on the
yieldstress of the crystal matrix and on the thermal expansion
difference between particle and matrix.   Figure 3.18 shows an
example of dislocation generation around a carbon-rich inclusion in
CdTe (Vere et al, 1985).   The observation of a star pattern is a
relatively rare event in this material as the matrix yield strength
is very low and dislocation propagation is relatively easy.
Cross-slip, glide and dislocation interaction readily lead to the
replacement of the star pattern by a diffuse dislocation tangle.
Similar features frequently referred to as "grappes' have been
observed in InP and have been identified as dislocation clusters,
prismatically punched from voids or In-rich inclusions (Augustus and
Stirland, 1982; Cockayne et al, 1983).   It was also noted (Cockayne
et al, 1981) that the incidence of 'grappes' increased when InP was
pulled from melts encapsulated with B_2O_3 with a high moisture
content.   Such conditions lead to enhanced phosphorus loss and
corresponding indium-enrichment of the melt. Prismatic punching is
also found around second-phase tellurium inclusions in CdTe and
$Cd_xHg_{1-x}Te$ (Figure 3.19)

     The differing densities and distributions of dislocations in
different materials is obviously related to the formation and
migration energies of these defects.   Comparisons between materials
are not straightforward, however.   There is only a limited amount of
data available, frequently obtained under widely differing
conditions of strain rate and temperature, and in comparison with
metals, the data shows a much wider spread of values due to the
relatively brittle nature of the material and the possibilities of
inaccuracies introduced by crack propagation.   For these reasons
perhaps the best guide to the relative strengths of semiconductor
materials is to present the data as shown in Table 3.1 which

Table 3.1   A comparison of the strength of semiconductor material
            based on the temperature T_σ at which the yield stress
            $\sigma = 2$ Kg mm^{-2} for a strain rate $\varepsilon = 10^{-4}$sec^{-1} (courtesy
            of J B Mullin, RSRE)

Material	T_σ (°C)	Reference
HgTe	−30	Cole (1982)
CdTe	5-50*	Vere (Unpublished)
$Cd_{0.2}Hg_{0.8}Te$	20	Cole (1982)
InSb	285	Patel and Choudhari (1963) Bell and Willoughby (1966)
GaSb	465	Patel and Choudhari (1963)
Ge	515	Patel and Choudhari (1963)
GaAs	700	Laister and Jenkins (1973)
Si	815	Patel and Choudhari (1963)

*CdTe yield strength varies by a factor of two or more, depending
upon illumination level and carrier concentration.

indicates the temperature at which material deformed by bending at a
strain rate of 10^{-4}sec^{-1}shows a yield stress of 2 Kg mm^{-2}.

   The final density and distribution of dislocations and
precipitates in a particular material has been shown by Williams and
Vere (1986) to depend critically on the formation and migration
energy of the point defects, the cooling rate and the cooling
temperature interval (which determine the *supersaturation* of point
defects) and the mechanical properties of the lattice and the
external stresses (which determine the *amount* of dislocation
nucleation). The balance between point defect and dislocation
content determines the ratio of homogeneous to heterogeneous
nucleation of excess second-phase (eg Te in CdTe). If the point

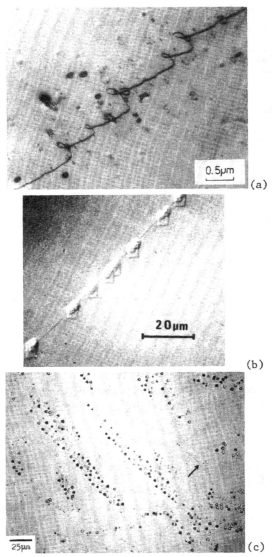

Fig. 3.20   Helical dislocation formation in (a) GaAs (courtesy of D
Stirland, Plessey Co), (b) InP (courtesy of G Brown,
RSRE) and (c) $Cd_{0.2}Hg_{0.8}Te$ (courtesy of D J Williams,
RSRE)

defect concentration is high and the incidence of dislocations is low, homogeneous precipitation occurs on a fine scale throughout the lattice, giving rise to small (<500A diameter) voids, {111} loops and precipitates. A high dislocation density provides an efficient sink for the mobile point defects leading to heterogeneous precipitation of excess second-phase Te at such sites.

Whilst precipitates and inclusions can generate dislocations, dislocation climb can also result in precipitation. Since the [110] dislocation in a polar lattice must contain a double half-plane to accommodate both anion and cation species, dislocation climb requires either the presence of an equal excess of both species (which is unlikely) or, as proposed by Petroff and Kimmerling (1976), the rejection of one species. The rejected atoms may be incorporated on the opposite sublattice, forming antisite defects or, as the concentration increases, may cluster to form dislocation loops or precipitates. Such "climb debris" is frequently found in the vicinity of heavily jogged dislocation lines.

In GaAs (Stirland et al, 1978), InP (Brown et al, 1983) and CdHgTe (Williams and Vere, 1986), climbing screw dislocations generate helices which break up to form dislocation loops lying on {110} planes (Figure 3.20). In the case of growth from In-rich or Te-rich melts, such loops are In-decorated in InP or Te-decorated in $Cd_xHg_{1-x}Te$, although it is not clear whether the decoration results from the climb process itself or from subsequent condensation of In or Te as a result of retrograde solid solubility. Precipitate decoration of the loops was also observed in GaAs although the stoichiometry of the material was unknown and the material was heavily Te-doped. The precipitates were too small for chemical analysis.

A further example of the complexity of the interaction between dislocations, native and impurity point-defects and precipitates is the occurrence and distribution of the EL2 deep donor level in GaAs (Martin, 1981) which compensates residual shallow acceptors such as carbon to produce semi-insulating material. Although the EL2 level is enhanced in regions of high dislocation density (Holmes et al, 1983; Matsumura et al, 1983; Nanishi et al, 1983; Skolnick et al, 1984 and Barnett et al, 1985) its presence in dislocation-free regions - albeit at a much lower level - suggests that it is not solely attributable to the dislocation. One possible explanation of the EL2 defect is that it is an As_{Ga} antisite defect (Weber et al, 1982; Lagowski et al, 1982; Zou Yuanxi, 1982), whose concentration is enhanced in the vicinity of dislocations due to an increased V_{Ga} concentration resulting either from preferential gettering of Ga to the dislocations or vacancy generation as a result of dislocation climb. More recently other mid-gap levels have been detected (Taniguchi and Ikoma, 1982, 1983; Lagowski et al, 1984). The very existence of these widely differing observations and the variability

of the electrical properties of the EL defect according to dopant
species suggests that these traps represent a generic defect
resulting from a complex dislocation native-defect impurity
interaction whose exact distribution and charge state is dependent
upon the thermal history and $|N_D-N_A|$ balance in the ingot.
Cathodoluminescence and photoluminescence studies provide increasing
evidence of this in both III-V (Dimitriadis, 1978; Salerno et al,
1981) and II-VI materials (Klimkiewicz and Auleytner, 1983;
Chamonal, 1982).

The observation that EL2 exists in dislocation-free material
suggests the possibility of precipitate/impurity interaction of the
type observed by Pautrat et al (1982) in II-VI compounds. They
observed that the tellurium precipitates in CdTe and ZnTe act as a
getter for impurities such as Cu and the alkali metals (and probably
others too), thus changing the electrical properties of the
surrounding matrix. Conversely, Schaake et al (1983) and Tregilgas
et al (1983) have shown that when $Cd_xHg_{1-x}Te$ is annealed in an Hg
atmosphere, the in-diffusion of Hg results in the dissolution of the
Te precipitate with consequent release of p-type impurities
previously locked in an electrically inactive state at the Te
precipitate.

3.5 Subgrain Boundaries

In II-VI compounds where weak interatomic bonding gives rise to
a high point-defect concentration and the generation of high
dislocation densities at low applied stress levels, dislocation
migration by climb and glide during cooling from high growth
temperatures leads to the formation of polygonised sub-grain
boundaries (Figure 3.6). Such boundaries have been observed in
many high-temperature grown II-VI materials, eg PbTe and $Pb_{1-x}Sn_xTe$,
CdTe, and $Cd_xHg_{1-x}Te$ (Mirsky and Shechtman, 1980; Schaake, 1983;
Williams, 1986; Williams and Vere, 1986).

Boundaries range in dislocation density from $<10^3 cm^{-1}$ to
$>10^8 cm^{-1}$ (corresponding to a misorientation θ ranging from <30 arc
secs to >5 arc mins), according to the mechanical properties and
thermal history of the material. At low densities the boundary
structure tends to resemble a loosely tangled dislocation array
spreading over several microns (Figure 3.21a), whereas at high
boundary-dislocation density (or after prolonged annealing), the
array forms an ordered network (Figure 3.21b).

In semiconductor device arrays these boundaries are
electrically active and particularly detrimental in terms of device
degradation since they can act as shorting paths between adjacent
active elements in the array. They have also been shown to act as
preferential sites for impurities (Vere et al, 1985) although it is
not yet clear whether the impurity gettering is an intrinsic feature
of the dislocation array or results from entrapment of impurities by

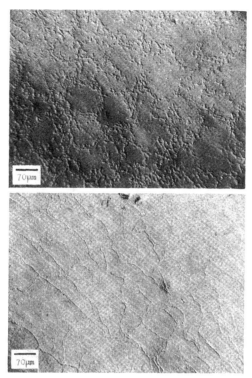

Fig. 3.21   Subgrain boundary in CdTe; a) diffuse boundaries
             containing approximately 10^5 dislocations cm^{-1};
             b) higher-angle linear boundaries containing $>10^8$
             dislocations cm^{-1} (courtesy of D J Williams, RSRE)

small Te precipitates decorating the array. A "denuded" zone devoid
of Te precipitates surrounds many such sub-grain boundaries. The
zone width is variable ranging from 1-10μm. In general the zone
exhibits n-type conductivity (except in regions around residual Te
precipitates lying on the boundary), indicating that the final
composition achieved by the local removal of Te from this region
lies on the Cd-rich side of stoichiometry. Similarly denuded zones
have been revealed by cathodoluminescence in the vicinity of
sub-grain boundaries in GaAs (Warwick and Brown, 1985).

3.6   Reduction of Defect Concentration

       It is evident from the work described above that the final
density and distribution of defects in semiconductor materials

Table 3.2 Crystal Defects - formation temperature ranges
and interactions

Melting Point	Twinning	Grain Boundaries	Point Defects	Disloc-ations	Impurities
0.9				Climb and glide	Vacancy-Impurity Complexes
0.8			Vacancies Divacancies Inter-stitials		Inter-stitial or substitu-tional atoms
0.7				Polygon-isation	
0.6			Vacancies + precip-itation		
0.5				Glide (under applied stress) (III-V)	
0.4					
0.3					
0.2					
0.1				Glide (under applied stress) (II-VI)	
Room Temp			Inter-stitials		Inter-stitials

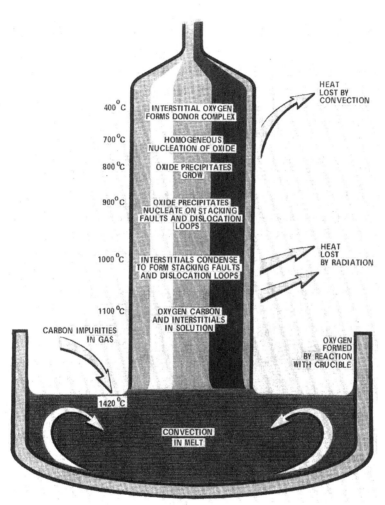

Fig. 3.22 Schematic diagram showing the defect formation processes
occurring during the growth and subsequent cooling of
CZ-silicon (courtesy of R Series, RSRE)

depends upon the types of defect interaction outlined in Table 3.2, which illustrates only the basic interactions and the approximate temperature ranges (or sequence) in which they might be expected to occur. The schematic diagram of Figure 3.22 shows the reactions occurring in CZ silicon during the growth and cooling processes. The complexity and interactive nature of the defect formation is further illustrated by the micrograph of Figure 3.23 which shows a rod-like defect in Fe-doped silicon. The exact details of the defect formation processes will depend upon:

    a)   The physical and mechanical properties of the material
    b)   Thermal gradients
    c)   Growth temperature
    d)   Impurity type and concentration

Attempts to reduce the defect density must therefore address one or more of these characteristics of the material or growth system.

    Although the scope for changing the materials properties is limited by device requirements (band-gap, electron mobility etc), alloying additions have been used to harden the lattice of several semiconductor materials. This solution hardening increases the

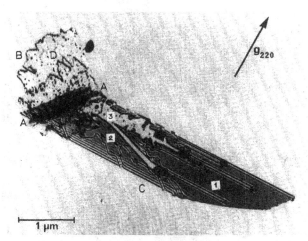

Fig. 3.23 Complex defect interactions in a p-n diode formed by B-diffusion into 1 Ωcm n-type (P-doped) CZ silicon. The defect AA is a rod-like precipitate of $FeSi_2$ which has nucleated prismatically punched dislocations B and a multiple stacking-fault sequence C1-3. Small subsidiary precipitates D, thought to be copper silicide, decorate the region of the dislocation array (after Cullis and Katz, 1974). Reprinted with permission from Phil.Mag. published by Taylor and Francis Ltd., London.

lattice friction (Peierls force), reducing point defect and dislocation migration. Examples of this approach include the use of In
additions to GaAs (Jacobs et al, 1983), Ge additions to InP (Seki et
al, 1976, 1978) and Zn additions in CdTe (Bell and Sen, 1985) and N_2
additions in Si (Abe et al, 1981).

In many cases alloying is used in conjunction with growth under
reduced temperature gradients to reduce thermal stress.
Several groups of research workers have published details of work
aimed at reducing temperature gradients in bulk growth processes,

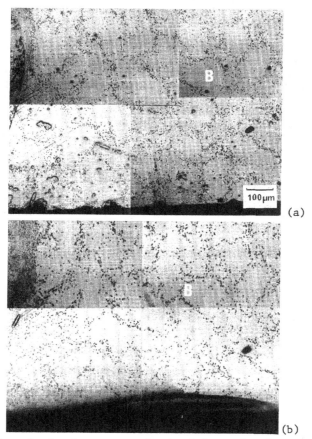

(a)

(b)

Fig. 3.24   Correlation between etch pit defects in a) CdTe epilayer
            and b) the underlying CdTe substrate (after Astles et al,
            1984).   Reprinted with permission from J.Electron.Mater.
            Copyright the Metallurgical Society, Warrendale, USA.

particularly in Czochralski growth (see Chapter 4). This work is frequently based on the use of suitable after-heaters to maintain ingot-temperature in the cooling region above the melt and, in the case of growth of volatile compounds such as GaAs by LEC (Liquid Encapsulated Czochralski), by careful control of the B_2O_3 encapsulant depth. In Bridgman growth, attention must be paid to the elimination of stress raisers (eg variability in ampoule bore) and growth ampoules are frequently carbon-coated to reduce sticking. The latter technique is often applied in the growth of CdTe, where residual oxygen in the sealed ampoule leads to enhanced reaction between the CdTe and the quartz ampoule and consequent adhesion during cooling of the solidified ingot. Soft-mould techniques involving the use of deformable linings have also been used with some success.

One of the most effective ways of reducing the defect concentration in a semiconductor material is obviously to reduce the growth temperature. Not only is the equilibrium concentration of point defects reduced, but so too is their mobility and hence their ability to interact to produce extended defects. Furthermore, the applied stresses occurring during cooling from the growth temperature to room temperature are also reduced. Unfortunately, low-temperature growth requires the use of a melt-grown substrate material.

3.7 Substrate-Layer Interface Defects in Epitaxy

In view of the need for large area deposition, the substrate will usually have been produced by high temperature bulk-growth (Bridgman or Czochralski) and will contain some or all of the defects discussed above. Defects associated with the layer/substrate interface include

a) Propagation of "threading" dislocations from the substrate into the layer

b) Propagation of surface damage resulting from improper preparation of the substrate

c) Nucleation of dislocations and stacking faults at particulate or precipitate residues on the substrate surface

d) Generation of dislocations due to lattice mismatch

e) Polycrystalline interfacial layers

f) Antiphase domains

g) Diffusion of impurities from the substrate

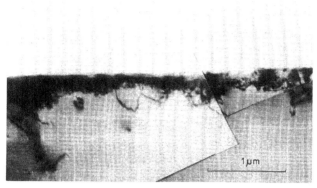

Fig. 3.25   Transmission electron micrograph showing surface damage
in CdTe polished with 0.5μm alumina powder (courtesy of M
Lunn and P Dobson, University of Birmingham)

In Chapter 2 we discussed the action of emergent dislocations
at the growth surface as nucleation sites for island growth. This
source of dislocations in the epilayer is particularly important
where dislocation densities in the substrate are high. Figure 3.24
shows the direct correlation between the CdTe substrate dislocation
structure and the deposited CdTe epilayer (Astles, 1984).

The dislocation density at the substrate surface may be
artificially enhanced by incorrect preparation procedures as
illustrated by Figure 3.25 where surface damage has been introduced
by polishing with 0.5μm particle size alumina (Lunn, 1986). In the
most extreme cases particulate residues, frequently of high Al and
Si content, remain on the surface after polishing and are a likely
source of the impurity concentration effects (ICE) observed by
Mullin et al (1983) and Tunnicliffe et al (1984), although it is
also possible that impurities present in the bulk or epilayer
material can be gettered to dislocation arrays during heat
treatment. Astles et al (1986) have shown that the use of a
"wash-melt" prior to LPE deposition can eliminate such interface
impurity spikes (Figure 3.26). A further source of interfacial
defects is the presence of residual oxides on the substrate surface
(Chew et al, 1985), whilst in alloy semiconductors similar stacking
faults and twins are observed to nucleate at free-metal globules on
the substrate surface, produced by dissociation during heat-cleaning
or ion-cleaning. Figure 3.27 shows the heavily-faulted structure
produced at the CdTe/InSb interface due to In droplets left on the
InSb substrate surface after ion-cleaning (Williams et al, 1985).

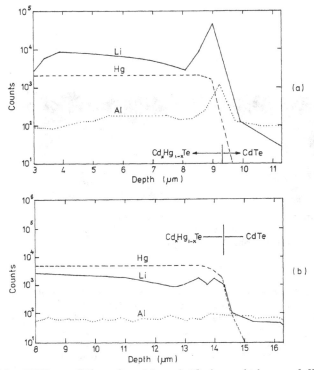

Fig. 3.26   SIMS profiles for Li and Al impurities and Hg in an
           LPE-grown $Cd_xHg_{1-x}Te$ layer on a CdTe substrate
           a) conventional growth without a "wash"-melt
           b) growth using a bismuth wash-melt (after Astles et al,
           1987).   Reprinted with permission from J.Electron.Mater.
           Copyright the Metallurgical Society, Warrendale, USA.

     In theory, the minimum dislocation density at the interface is
obtained when the sole remaining source is the misfit strain arising
from the lattice mismatch between layer and substrate.   In a
bicrystal whose individual lattice parameters are a_L (layer) and a_S
substrate ($a_L > a_S$) the misfit f is given by:

$$f = \frac{2(a_L - a_S)}{a_L + a_S} \qquad (3.4)$$

Table 3.3 gives the lattice parameters of several commonly-
encountered layer and substrate materials.   For the GaAs/CdTe system

Table 3.3   Lattice parameters (300K) of some elements and
            compounds in common use in semiconductor processing
            (a_O cubic except where stated)

Element or Compound		Lattice Parameter (Å)
Ni		3.52
Pd		3.881
Pt		3.923
Al		4.049
Au		4.079
Ag		4.08
TiO_2 (Tet)	c_O=2.9592	4.5933
Al_2O_3 (Trig)	c_O=12.991	4.759
SiO_2 (Hex)	c_O=5.4053	4.9133
CaF_2		5.4626
SrF_2		5.7996
BaF_2		6.2000
Si_3N_4 (Hex)	c_O=2.9075	6.6044
$MgAl_2O_4$		8.0831
$Y_3Al_5O_{12}$		12.01
$Gd_3Ga_5O_{12}$		12.376
Si		5.4307
ZnS		5.43
GaP		5.447
Ge		6.65
GaAs		5.6534
ZnSe		5.669
CdS		5.832
InP		5.8688
InAs		6.0585
GaSb		6.0955
HgTe		6.457
PbTe		6.460
InSb		6.479
CdTe		6.481

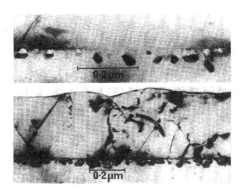

Fig. 3.27   Cross-section transmission electron micrograph of the
            interface between an MBE-grown CdTe epilayer and the InSb
            substrate (T=280°C). The dark particles below the
            interface in the upper micrograph are In droplets which
            have been generated on the InSb surface on heating the
            In-rich surface layer produced by ion-cleaning. The
            lighter areas are voids generated by the migrating
            droplets. The lower micrograph is a lower magnification
            view showing the full thickness of the epilayer (after
            Williams et al, 1985). Reprinted with permission from
            J.Vac.Sci.Technol. published by the American Institute of
            Physics, New York.

Fig. 3.28   Lattice image of the interface region of a (100) CdTe
            epilayer grown onto a (100) GaAs substrate. The white
            diffuse spots are mismatch dislocations at approximately
            seven lattice-plane spacings to accommodate the 14%
            mismatch between the CdTe (upper lattice) and GaAs (lower
            lattice) (courtesy of M Lunn, University of Birmingham).

a_S(GaAs) = 5.6534Å and a_L(CdTe) = 6.481Å giving an f value of 0.136 so that accommodation of mismatch strain by dislocations requires the insertion of a dislocation every seven lattice planes. Figure 3.28 shows the TEM lattice image obtained from such a GaAs/CdTe interface, confirming this prediction.

The nature and density of the dislocation array formed at the interface depends upon many factors including the mismatch strain, the growth temperature, the difference in thermal expansion coefficients between substrate and layer materials and their relative mechanical strengths. At high densities and low homologous temperatures (ie relative to the melting-point) the dislocations form a linear two-dimensional array, giving rise to a "cross-hatched" pattern at the interface; for lower densities and higher temperatures redistribution of the array occurs (Matthews, 1970; James and Stoller, 1984) as shown in Figure 3.29. At higher temperatures, cross-slip and climb of the dislocation array occurs and the dislocations are no longer confined to a planar array but show an exponential decrease in density with increasing distance from the interface in the softer material.

Where large lattice parameter mismatch must be accommodated in alloy systems, the required dislocation density can often be minimised by grading the interface composition by sequential growth of slightly mismatched composition. For example, the beneficial use of multiple $In_xGa_{1-x}P$ buffer layers of gradually increasing x value

Fig. 3.29   Transmission electron micrograph (plan view) of the
            dislocation array generated at the interface between a
            CdHgTe epilayer and a CdTe substrate (courtesy of M
            Lyster and R Booker, Oxford University, and N Chew, RSRE)

has been demonstrated by Olsen (1975) in the growth of InGaP layers on GaP substrates.

Although equation 3.4 can be used to predict the minimum interfacial dislocation density in low yield-stress systems, this prediction implies that the only accommodation mechanism is dislocation generation. In 'hard' systems part of the strain is relieved by <u>elastic</u> distortion of the weaker lattice. The elastic strain ϵ increases as the layer thickness h increases until the yield stress of the material is reached and the dislocation array forms. According to Matthews (1979) the critical thickness h_c at which this occurs is given by:

$$h_c = \frac{G_S(f - \epsilon)b}{2\pi(1+v)\epsilon^2(G_L+G_S)} \left[\ln \frac{R}{b} + 1\right] \qquad (3.5)$$

where G_S and G_L are the shear moduli of the substrate and layer respectively, R is the radius of distortion due to the dislocation, b its Burgers vector and v the Poisson ratio. For epilayer thicknesses less than h_c dislocation-free interfaces are possible but usually of limited value since, in practice, the elastic strains generated in the layer are rarely isotropic.

So far we have considered only macroscopically isotropic strains. Recent TEM evidence suggests, however, that once the lattice mismatch exceeds a few percent, localised strain relief may occur by formation of 'single-crystal' domains of differing orientations. In the growth of fluorides on silicon, for example, Phillips and Gibson (1984) showed that for mismatches exceeding 7% two domains occurred simultaneously:

TYPE A: $(111)_{CaF_2}$ $\parallel$ $(111)_{Si}$: $[110]_{CaF_2}$ $\parallel$ $[110]_{Si}$

TYPE B: $(111)_{CaF_2}$ $\parallel$ $(111)_{Si}$: $[110]_{CaF_2}$ $\parallel$ $[114]_{Si}$

For fluorides with lower mismatch values one or other of these occurs exclusively.

In polar/non-polar semiconductor combinations (eg (100)GaAs $\parallel$(100)Ge it is also possible to generate antiphase domains at the interface (Pond et al, 1984). These defects are revealed by contrast reversal on plan-view TEM. The suggested origin is a change in lattice stacking sequence from

Ge (-1/4, -1/4, -1/4)— Ga (0,0,0) — As (1/4, 1/4, 1/4)

to

Ge (-1/4, -1/4, -1/4) — As (0,0,0) — Ga (1/4, 1/4, 1/4)

in adjacent areas of the growth surface due to odd-number atomic
step heights on the substrate surface. Their occurrence is limited
to {hk0} orientations. Similar features have been observed in
(100)GaAs ||(100)Si by Wang (1984), who showed that their formation
could be suppressed by exposing the Si surface to an As flux to form
a Si-As bond giving surface steps of even-numbered atomic heights.

Provided that the interfacial lattice mismatch is relieved
solely by dislocation generation, the vertical extent of the
dislocation array reduces with reducing growth temperature since
movement of the dislocations from their point of origin by
cross-slip and climb becomes progressively more difficult. At the
same time, the potential for growth defect annihilation by surface
migration of adatoms becomes progressively more difficult too, so
that correct surface preparation, stoichiometry and reconstruction
become increasingly important for the achievement of good crystal
growth under kinetically-limited conditions. This is particularly
so for heteroepitaxial structures where the lattice parameters and
chemical compositions of layer and substrate are quite different.
Much of the recent progress on systems such as silicon on sapphire
(SOS) (Cullen, 1978; 1983), GaAs on silicon (Fischer et al, 1984;
Akiiyama et al, 1984; Wang, 1984) and CdTe and $Cd_xHg_{1-x}Te$ on
sapphire (Tennant, 1984; Hyliands, 1986) and $CdTe/Cd_xHg_{1-x}Te$ on GaAs
(Nishitani et al, 1983; Giess et al, 1985) has been attributable to
the combination of low-temperature growth and progress in substrate
preparation and characterisation.

CHAPTER 4

GROWTH FROM THE LIQUID PHASE

4.1 Introduction

     There have been many reviews of liquid-phase growth of
semiconductor materials (Brice, 1965, 1973, 1986; Elwell and
Scheel, 1975) and whilst it is instructive to begin with a brief
review of the basic techniques, this chapter is intended to
emphasise the similarities underlying the apparent wide diversity of
techniques. It also focuses attention on recent developments in
Czochralski (CZ) and Bridgman growth, since these techniques are
increasingly important in commercial production of bulk devices and
the large-area substrates for epitaxial growth.

     Liquid-phase growth can be sub-divided into growth from the
melt and growth from solution. In melt growth the driving force for
solidification is the movement of an imposed normal temperature
gradient relative to the liquid-solid interface. The factors
governing growth are therefore the relative rate of movement of the
gradient and interface (ie the "pull rate") and the steepness of the
temperature gradient. If the gradient is too shallow,
constitutional super-cooling will cause breakdown of the planar
liquid-solid interface to give a cellular or dendritic structure as
discussed in Chapter 3. In systems where the melt is contained in
an ampoule, spurious nucleation may also occur on the ampoule walls
ahead of the advancing interface, producing a polycrystalline
ingot. If the imposed thermal gradient normal to the interface is
too steep, convection in the melt leads to non-uniform impurity
distribution and to non-stoichiometry in alloy materials. A steep
gradient also imposes severe lattice strains on the solidified
material, leading to dislocation generation or cracking. The art of
good crystal growth is, therefore, to balance the pull rate against
the rate of heat loss from the interface and to reduce the applied

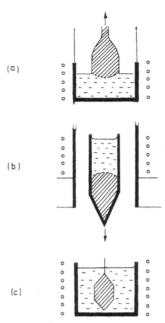

Fig.4.1 Liquid-phase bulk-growth techniques: a) Czochralski,
         b) Bridgman, c) solution growth.

temperature gradient to the minimum consistent with interface
stability and growth control. The axial temperature gradient G,
required to suppress constitutional supercooling, is given by the
equation

$$\frac{G}{R} \geq \frac{mC_o}{Dk_o} \left[1 - k_o\right]$$

(4.1)

where R is the growth rate, m the slope of the liquidus dT/dC with
respect to the dominant alloy component or impurity, C_o the initial
melt concentration of that component and k_o its segregation
coefficient. The upper limit to the temperature gradient is
determined by the mechanical properties of the material to be grown
and the type of device to be fabricated, since the latter dictates
the acceptable crystal quality. Whilst high applied gradients can
be easily tolerated in oxide materials, where atomic bond strengths
are high and defect generation difficult, in II-VI materials,
applied strain is rapidly dissipated by high-density dislocation
generation. Recent developments in III-V and II-VI semiconductor
growth have therefore been directed towards reduction of applied
gradients, in combination with lattice-hardening by the use of
alloying additions.

Even within the acceptable range, the temperature gradients in the growth-system produce convective fluid flow within the melt and the vapour phase above it. The melt convection currents play an important part in determining the shape and stability of the growth surface, the erosion rate of the crucible or ampoule material and its transport to the growth surface and these effects will be examined in more detail in Sections 4.3 to 4.6.

In solution growth the liquid and solid phases are practically isothermal and thermal convection is minimal. The liquid is slowly cooled to produce a super-saturation of the required crystallising phase, which then precipitates onto the surface of a suitable single-crystal "seed". In this case the driving force for crystallisation is the concentration gradient established in the melt above the growth surface. The crystallising phase is then conveyed to the growth surface by "solutal" convection and the art of good crystal growth is to ensure that the supply rate is constant and matched to the rate at which the adsorbed atoms can be accommodated at the correct lattice sites on the growth surface.

4.2 The Basic Liquid-Phase Growth Techniques

Liquid phase growth embraces such diverse techniques as the Czochralski and Bridgman bulk-growth processes and growth from fluxes and solutions. The basic similarities of melt and solution growth processes are illustrated schematically in Figure 4.1. Growth from the melt (Figures 4.1a and 4.1b) is appropriate where the material melts congruently and can be solidified in single-crystal form by the imposition of a temperature gradient, T_1-T_2, where the melting point of the material lies between T_1 and T_2.

The vertical Bridgman technique, in which the melt is contained in a growth ampoule and is solidified by lowering the ampoule through the temperature gradient, is a simple, low-cost technique with the added advantage that the melt temperature increases with distance from the solid-liquid interface and the system is therefore density-stable and less prone to convection effects than the Czochralski technique, though radial gradients ensure that convection cannot be eliminated entirely. A disadvantage is that the low growth rates used, typically 0.1-1.0 cm hr^{-1}, mean that the melt is in contact with the ampoule for long periods, increasing the risk of impurity pick-up. A second disadvantage is that adhesion of the solid material to the ampoule wall or compression of the solid by the contracting ampoule during cooling can lead to the development of stresses high enough to nucleate dislocations in the material. A further drawback of the technique is that the ampoule wall acts as a preferential, spurious nucleation site, resulting in polycrystalline rather than single-crystal growth unless the temperature gradient and liquid-solid interface shape are

well-controlled. Horizontal versions of the Bridgman technique are
also employed but growth control is complicated by difficulties in
controlling interface shape. These are attributable to asymmetry
in the thermal field and the presence of a relatively large area of
melt in contact with a convecting gas phase.

Advantages of Czochralski growth lie primarily in the lack of
direct contact with the crucible and the use of faster growth rates
than are possible with flux or solution techniques. The diameter of
the growing ingot is governed by the meniscus height, which
determines the optimum pull rate. Modern techniques of optical
interface- monitoring (Gross and Kersten, 1972) or continuous
crucible weighing (Zinnes et al, 1973; Bardsley et al, 1974, 1975)
have eliminated many of the earlier problems in diameter control.
The development of the liquid-encapsulated Czochralski (LEC)
technique (Mullin et al, 1970), in which the melt is surrounded by a
layer of molten B_2O_3 or similar encapsulant in a pressurised growth
chamber has extended the technique to the growth of volatile
materials such as GaAs and InP. Present-day problems centre around
the need to reduce the high dislocation-density characteristic of
CZ-grown III-V materials and to reduce the incorporation of
contaminants from the crucible or exposed hot metal parts of the
pulling system or heating element. The former is tackled by a
combination of alloy-hardening, temperature-gradient reduction and
improved diameter control, whilst the latter is tackled by the use
of forced convection or magnetic damping to reduce convective
mass-transport in the melt and the use of low-pressure or swept-gas
environments to prevent vapour transport of impurities to the
crystal or melt surface. The success, or failure, of such
techniques is discussed further in Section 4.6.

There are many other containerless techniques, such as floating
zone (FZ) and the Verneuille technique (Hurle, 1979), which are
particularly applicable for the growth of high melting point
materials, eg oxides. They are not discussed here because the basic
principles are similar to those of the CZ technique.

In solution-growth the required material is dissolved in a
liquid which is then evaporated or cooled to produce
supersaturation. Flux growth is a sub-division of solution growth
in which the "flux" into which the required materials dissolved is a
"mother-liquor" of molten salts, oxides, fluorides etc of lower
melting point than the required material. Crystal growth from high
temperature solutions has been extensively reviewed by Elwell and
Scheel (1975). For a recent survey of flux growth see Wanklyn
(1983). In semiconductor technology, bulk solution growth is
largely confined to the production of research quantities of new
materials (particularly those with high-temperature phase
transitions). Though cheap, the technique is difficult to seed,
growth is limited and flux inclusions or impurity striations   are

common. There are, however, some notable exceptions where solution
growth is the major production method. These include hydrothermally
grown crystalline quartz for optical and piezo electric applications
triglycene sulphate (TGS) crystals for infra-red detector
applications and ammonium dihydrogen phosphate (ADP) and potassium
dihydrogen phosphate (KDP) crystals for electro-optic applications.

One form of solution growth has, however, found widespread
application in semiconductor technology. This is liquid-phase
epitaxy (LPE), in which the required alloy material is dissolved in
one of its constituents, eg GaAs growth from Ga melts, InP from In
melts or $Cd_xHg_{1-x}Te$ from Te or Hg melts. The melt is cooled to
generate a supersaturation of the alloy phase, which is precipitated
in single-crystal layer form onto a suitable single-crystal
substrate. Layer thicknesses are typically 1-50 μm. Prolonged
controlled growth is difficult to sustain and for that reason melt
and substrate are usually brought together for a limited period and
then separated. This can be achieved by dipping, tipping or sliding
techniques, as illustrated schematically in Figure 4.2. Despite
their seeming diversity, all these liquid-phase techniques have
several common features. In each system the liquid phase is in
contact with a nucleating site at which solid material is deposited. This takes the form of either a seed crystal, on which
deposition occurs on all exposed crystallographic faces, or a
substrate, of which only one face (of known crystallographic

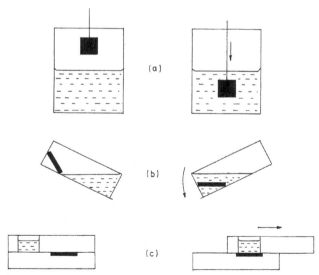

Fig.4.2 Liquid-phase epitaxial (LPE) techniques: a) dipping
         b) tipping, c) sliding-boat.

orientation) is exposed for growth. Thus each system contains a
liquid-solid interface at which growth (or dissolution) is
determined by the heat-flow and concentration gradient in the
immediate vicinity. A second common feature is the presence of the
gas phase. In Bridgman and solution growth it is usually remote
from the liquid-solid interface and generally is only directly
involved in the solidification of high-vapour-pressure materials,
where the equilibrium liquid-phase concentration is dependent upon
the gas-phase temperature. In Czochralski growth, however, the
thermal properties of the gas-phase play a more direct part in
determining the shape and stability of the liquid-solid interface
and the optimum growth rate.

4.3 Convection in Melts

When a melt is heated from below, the buoyancy forces generated
cause the hotter, less dense liquid, to rise and the cooler liquid
to fall, generating a convective flow. Convective flow may also
result from the gradient of surface tension forces (Marangoni
convection) or gradients in solute or alloy concentration (solutal
convection). The resulting flow patterns and their effect on
temperature distributions (and hence liquid-solid interface shape

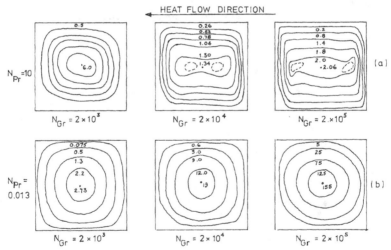

Fig.4.3   Stream-function contours generated in fluids with
          Prandtl numbers of 0.013 and 10 which are contained
          in square enclosures heated and cooled at the vertical
          sides under an increasing horizontal temperature
          difference (after Stewart and Weinberg, 1972).
          Reprinted with permission from J.Cryst.Growth published
          by North Holland, Amsterdam.

and mass transport) have been extensively studied (Carruthers, 1976; Hurle, 1976, 1983; Normand et al, 1977).

The important parameters governing the onset and nature of the convection patterns in melts are the thermal characteristics of the material, the aspect ratio of the container and the imposed temperature gradients. The Prandtl number N_{Pr} is one of the most important factors governing the fluid flow patterns and is defined as the ratio ν/κ where ν is the kinematic viscosity of the liquid and κ is the thermal diffusivity. It is dependent only on the thermal characteristics of the liquid and independent of container geometry or applied temperature gradient. In high Prandtl number systems ($N_{Pr} \geqslant 1$) such as oxides, strong fluid flow is largely confined to boundary layer regions (Figure 4.3a) whereas in low Prandtl number materials such as metals and semiconductors the streamlines are more generally distributed throughout the bulk but stream velocities are much higher under high applied temperature gradients.

The temperature gradient provides the driving force for the convection. In horizontal systems this driving force is characterised by the Grashof number Gr where

$$Gr = \alpha g l^4 \Delta T_H / \nu^2 \qquad (4.2)$$

in which g is the gravitational constant, α the thermal expansion coefficient for the melt, ΔT_H is the imposed horizontal temperature difference over the length l separating the opposite boundaries of the melt and ν is the kinematic viscosity of the liquid. In vertically-driven convection systems the relevant dimensionless number is the Rayleigh number Ra defined by

$$Ra = \alpha g d^4 \Delta T_V / \nu \kappa \qquad (4.3)$$

where ΔT_V is the vertical temperature gradient. Further dimensionless numbers appropriate to fluid flow are summarised in Table 4.1. Their definitions and effects on fluid flow are discussed in detail in the review of convective instability given by Normand et al (1977).

As the Rayleigh (or Grashof number) increases with increase in ΔT the fluid begins to move by convective flow and further increases lead to progressive breakdown into smaller convection cells as shown in Figure 4.4. The onset of each Mode is governed by a series of eigenvalues $Ra^{i,n}$ where i is the mode number and n the mode symmetry. In practice only n=1, the asymmetric mode with flow up one side of the container and down the other, or n=0 the axisymmetric mode, are of physical significance. Since Ra $\propto$ d^4 however, i can attain fairly large values in cylindrical containers with large d/a aspect ratio, eg typical Bridgman growth ampoules.

Table 4.1 Some Dimensionless Numbers in Fluid Flow

Prandtl	$Pr = \nu/\kappa$
Peclet	$Pe = Vl/\kappa$
Reynolds	$Re = Vl^2/\nu$
Grashof	$Gr = \alpha gl^4 \Delta T_H/\nu^2$
Rayleigh	$Ra = \alpha gd^4 \Delta T_V/\nu\kappa$
Schmidt	$Sc = \nu/D$
Marangoni	$Ma = \theta\Delta Tl/\rho\nu\kappa$
Nusselt	$Nu = Ql/k\Delta T$
Biot	$Bi = Ql/k$
Stanton	$St = Q/\rho C_p V$
Graetz	$Gz = \dot{m}C_p/kl$

ν	kinematic viscosity	α	thermal expansion coefficient
κ	thermal diffusivity	ΔT_H	temperature difference - horizontal
k	thermal conductivity	ΔT_V	temperature difference - vertical
V	flow velocity	C_p	specific heat - const.press.
d	fluid depth	ζ	surface tension
l	characteristic length	θ	temp.coeff. of ζ
Q	heat flux	ρ	fluid density
D	mechanical diffusivity	g	gravitational constant
$\dot{m}$	rate of mass flow		

Further increase in Ra or Gr leads to the development of
instability in the flow pattern, usually taking the form of
oscillatory motion of the convection cells within the melt, the
oscillation frequency increasing as Ra increases. The exact
mechanism by which the increasing frequency and diversity of
convection modes leads to turbulence is the subject of much debate,
but is of somewhat academic interest to the practical crystal grower
since the temperature gradients required are usually far in excess
of those yielding good quality single-crystals. Practical gradients
do, however, result in Ra numbers characteristic of non-stable
convective modes, which can result in adverse effects such as
periodic melt-back of the growth interface and the non-uniform
incorporation of alloy components or impurities to produce the
growth striae discussed in Chapter 3. For this reason practical
growth systems employ crystal and/or crucible rotation to generate
thermal symmetry and to impose a dominant 'forced convection' flow
pattern through the action of centrifugal force. In general this
stable pattern is preferable to the thermal convection pattern, in
controlling liquid-solid interface stability but the increased flow
velocities may lead to increased crucible erosion rates and higher
impurity mass-transport efficiency.

4.4 Liquid-Solid Interface Shape

The shape of the growth surface is dependent upon its
crystallographic orientation and the local temperature gradient.
The latter is determined by the external gradients and the way in
which they are modified by the thermal characteristics of the melt,
particularly convection. Figure 4.5 illustrates three possible
schematic interface shapes, defined in terms of the magnitude and
sign of d', the depth of the centre of the interface with respect to
its edge. The diagram is drawn for the case of CZ growth but

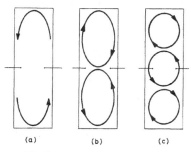

(a)   (b)   (c)

Fig.4.4   Convection cell formation as a function of increasing
          Rayleigh number: a) Ra = 230, b) Ra = 280 C) Ra = 382
          (after Olsen and Rosenberger, 1979). Reprinted with
          permission from J.Fluid Mechanics published by Cambridge
          University Press, Cambridge, England.

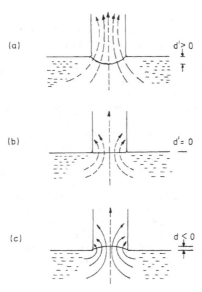

Fig.4.5 The relation between liquid-solid interface shape
and heat flow across the interface. a) convex,
b) flat, c) concave. The interface becomes progressively
more concave as heat loss via the solidified material
decreases or the centre to edge temperature differential
in the crucible decreases.

applies equally to Bridgman growth. The simplest and arguably the
best case from the point of view of good quality growth is the
planar interface, $d' = 0$. Unfortunately the condition is difficult
to maintain over long periods in the changing thermal environment
during growth and slight thermal fluctuations can generate an
interface with $d'<0$, ie concave with respect to the solid.

Although not necessarily unstable, the concave interface is
indicative of an excess power input raising the melt temperature.
In growth of oxides and other materials with high surface tension,
the excess heat input can move the liquid-solid interface several
millimetres above the plane of the melt surface. Radial heat losses
then become important in determining the exact interface shape at
the periphery of the crystal and, in extreme cases, ring facets and
'spines' (Figure 4.6) can develop. The spines are a result of rapid
radial heat loss and hence rapid solidification on facet planes
parallel to the growth axis. The ring facet occurs when the melt
isotherm lies parallel to a facet plane normal to the growth axis.
The mechanisms by which such defects form is illustrated
schematically in Figure 4.7. The concave interface also leads to

Fig.4.6   a) Plan view of the top surface of a CZ-grown ingot of $Ba_2NaNb_5O_{15}$ showing the development of large orthogonal [100] spines and smaller intermediate [110] spines. The triangular central feature is the original seed crystal (courtesy of B.Cockayne, RSRE, Malvern). b) Etched cross-section showing the ferroelectric domain structure characteristic of the formation of an annular or "ring" facet on the (001) growth surface. The small white features in the central area are dislocation arrays propagating from the original seed crystal.

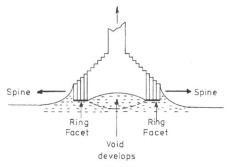

Fig.4.7   Schematic illustration of spine formation due to rapid lateral heat loss from the (100) facet planes. At the periphery of the crystal, the melting isotherm lies parallel to the (001) facet plane, producing an annular or "ring" facet.

entrapment of bubbles on the central region of the interface.
Again, this is a particular feature of high surface tension
materials and in extreme instances can lead to the propagation of
elongated voids in the material.

From this analysis we conclude that the optimum interface shape
for CZ growth is slightly convex, since under these conditions the
inevitable slight power and temperature fluctuations do not result
in unstable growth. If, however, d' becomes too large, close-
packed, high-index crystallographic planes steeply inclined to
the growth axis then lie parallel to the solidification isotherm and
facets can form on the growth surface. This is observed in the
growth of sapphire and garnet crystals, where the high transparency
of the crystal to radiation from the melt causes efficient heat
extraction via the crystal producing a conical section interface.
This gives rise to three {111} facets in the central region and the
resulting segregation of dopants or impurities generates a strained
central core in the ingot (Figure 4.8) (Cockayne and Roslington,
1973).

In Bridgman-growth the concave interface, whilst not unstable,
generally results in poor crystallinity since it promotes
'edge-grain' nucleation at the wall of the growth ampoule, whereas a
slightly convex interface tends to occlude such growth and, as in
Czochralski growth, is the more favourable situation. In view of
the importance of interface shape in determining crystal quality,
much effort has been devoted to the construction of thermal models
of the heat-flow in Bridgman and Czochralski systems.

1 cm

Fig.4.8   Central {211} and peripheral {110} facet formation on
           the liquid-solid interface of a [111] axis $Gd_3Ga_5O_{12}$
           (GGG) crystal. The facets are revealed by Twyman-Green
           interferometry due to their surrounding strain fields
           (after Cockayne and Roslington, 1973). Reprinted with
           permission from J.Mater.Sci. published by Chapman and
           Hall, London.

4.5 Thermal Modelling of Bridgman Systems

Chang and Wilcox (1974) solved the steady state differential equation for temperature T in an infinitely long solid cylindrical rod at the axis of a heater/cooler configuration (an idealised vertical Bridgman system):

$$\frac{k}{\rho C_p} \left[\frac{\partial^2 T}{\partial r^2} + \frac{1}{r} \frac{\partial T}{\partial r} + \frac{\partial^2 T}{\partial z^2} \right]^2 - \frac{V \partial T}{\partial z} = 0 \qquad (4.4)$$

where k is the thermal conductivity, ρ the density, C_p is the heat capacity, r is the radial position, z is the axial position measured downwards from the heater/cooler boundary and V is the growth rate (assumed equal to the rate of downward movement of the rod). The resulting curves of interface shape are shown in Figure 4.9 for differing values of $\varphi = T\text{-}T_c/T_h\text{-}T_c$ where T_h and T_c are the heater and cooler temperatures respectively and for Peclet numbers of 0.1, 1.0 and 2.0 (high for low-conductivity materials and high growth rates).

In more recent work Jones et al (1982, 1984) have used an electrical analogue model to study the temperature distribution in the Bridgman system used in the growth of $Cd_xHg_{1-x}Te$. This work has emphasised the important effects of heat loss from the base of the

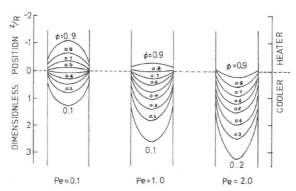

Fig.4.9 Relationship between calculated isotherm shape and rate of ampoule movement (increasing Peclet number). $\phi = (T_1\text{-}T_c)/(T_h\text{-}T_c)$, where T_1 is the interface temperature, T_h the heater temperature and T_c the cooler temperature. Curves neglect the influence of latent heat and are for a Biot number H = 2.0 (after Chang and Wilcox, 1974). Reprinted with permission from J.Cryst.Growth published by North Holland, Amsterdam.

ampoule in establishing the interface shape during the early stages
of growth. It also considers the effects of the temperature
dependence of thermal conductivity and the thermal transparency of
the silica ampoule. Figure 4.10 shows the temperature difference
between centre and edge (T_C and T_E respectively) as a function of
axial position. The solid line is computed using a step function
change in thermal conductivity from 1.4 $wm^{-1}C^{-1}$ for the solid to 10
$wm^{-1}C^{-1}$ for the melt, whilst the dotted line is plotted for the case
of a linear increase from 1 $wm^{-1}C^{-1}$ at 700°C to 5 $wm^{-1}C^{-1}$ at 820°C.
Ramp functions for the change in thermal conductivity generate
curves lying between these extremes. For practical conditions, some
increase in thermal conductivity on melting is to be expected, so
that the general observation is of a concave isotherm (and hence
interface shape) at the tip of the ingot. This leads to increased
nucleation at the ampoule wall compared with the central region of
the interface and the greater possibility of polycrystalline
growth. Although the melting-point isotherm flattens as the length
of solidified ingot increases, it may be difficult to remove the
grain boundaries already established. The shape of the isotherm in
the early stages of growth is therefore critical. The earlier work

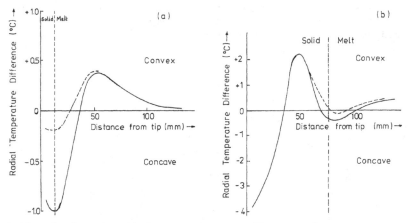

Fig.4.10 Calculated radial temperature difference ($T_{edge} - T_{centre}$)
as a function of distance from the top of the growth
ampoule a) after 10 mm growth of solid material; b) after
80mm growth of solid material.
_____ Thermal conductivity step-function change from
        1.0 $wm^{-1} C^{-1}$ to 10 $wm^{-1} C^{-1}$ on melting;
- - - - Thermal conductivity increasing linearly from
        1.0 $wm^{-1} C^{-1}$ at 700°C to 5 $wm^{-1} C^{-1}$ at 820°C (after
        Jones et al, 1984).
Reprinted with permission from J.Cryst.Growth published by
North Holland, Amsterdam.

by these authors showed that the magnitude of T_E-T_C could be increased by using a high-conductivity ampoule support stem at the base of the ingot. Errors in the choice of 'view-factors' used in the computation led to an over-estimate of the effect, however, and it is uncertain whether or not a convex interface shape can be produced in practical systems using a stem conductivity of 25 $wm^{-1}C^{-1}$. Nevertheless, the concavity of the interface can be significantly reduced.

In modelling vertical Bridgman systems, thermal convection is usually neglected since, in the idealised system at least, the melt temperature rises with increasing height above the liquid solid interface so that buoyancy-driven convection does not occur. In the Bridgman growth of $Cd_xHg_{1-x}Te$, however, the large segregation coefficient results in a build-up of Hg-rich material ahead of the interface. If the interface is concave, as is frequently the case in the initial stages of growth, the denser Hg-rich material can move towards the central region, promoting solutal convection and increasing the concavity (Capper, 1982). Capper et al. (1984) have shown that this effect can be countered

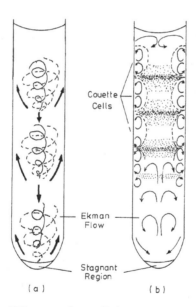

Fig.4.11 Schematic illustration of forced-convection in a Bridgman ampoule in the Accelerated Crucible Rotation technique (ACRT); a) accelerating, b) decelerating (after Capper et al, 1984). Reprinted with permission from J.Cryst.Growth published by North Holland, Amsterdam.

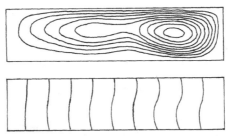

Fig.4.12 Typical streamlines and isotherms for Ra=500 and Pr=0.015
in a horizontal Bridgman boat shown in side elevation.
A linear temperature profile is imposed on the free
surface and at the bottom of the crucible (after
Crochet et al, 1983). Reprinted with permission from J.Cryst.
Growth published by North Holland, Amsterdam.

by using the Accelerated Crucible Rotation Technique (ACRT)
developed by Scheel and Schultz-Dubois (1971). The flows observed in a simulation experiment were extremely complex, comprising a strong Ekman flow (radial) in the immediate vicinity of
the liquid-solid interface and the formation of Couette convection
cells at the ampoule wall (Figure 4.11). The apparent effects of
these flows were that the Ekman flow suppressed the solutal
convection, whilst the Couette flow suppressed nucleation on the
ampoule walls resulting in improved crystallinity. The technique
can be generally applied in Bridgman systems, but care must be taken
to ensure that the dramatically increased flow rates do not lead to
increased contamination through an accelerated ampoule erosion rate.

Although attempts have been made to model the thermal profiles
in horizontal Bridgman boats (Hurle et al, 1974; Crochet et al,
1983), the situation is extremely complex due to the effects of
convection above the free surface of the melt and the frequent use
of D-section melts and containers with rounded end-faces. The
simple streamline pattern observed at low Rayleigh numbers (Figure
4.12) breaks down at approximately Ra > 500 corresponding to an
0.3°C temperature difference in GaAs over an axial boat length of
10cm. For practical growth systems with temperature gradients >10
°C cm^{-1} (Ra > 16,500) no steady-state solution exists and a periodic
flow pattern is established due to the superpositioning of a short
period oscillation on a longer one. The effect of this oscillatory
flow pattern on the liquid-solid interface is difficult to model.
Fig. 4.13 shows the calculated streamlines and interface shape at
one instant in the oscillation time period and its change with
increasing proportion of solidified material. Crochet et al
conclude from their study that the interface oscillates around a
mean position with a periodicity equal to that of the liquid-phase
flow, so that the interface is subject to periodic remelting.

4.6   Thermal Modelling of CZ Systems

In CZ growth the buoyancy-driven convective flows interact with the forced-convection imposed by the rotation of the seed crystal and/or crucible.   In addition to the problems of the inherent complexity of the situation, the thermal modelling is further complicated by the change in crystal length and diameter with time (with associated changes in melt depth, position of both crystal and melt with respect to the heat source and, in consequence, a continuously changing thermal distribution in both crystal and melt).

The onset of forced-convection due to rotation is marked by the appearance of a Taylor-Proudman column induced by the action of the centrifugal forces from crystal and crucible rotation, causing displacement of material in the surface region towards the crucible wall with a consequent upward central flow.   It is therefore dependent upon several factors, including crystal and crucible rotation rates, the aspect ratio of the melt and the viscosity and surface tension of the liquid melt (ie the efficiency of coupling between the crystal and melt).   Practical determination of the rotation rate required as a function of diameter and melt viscosity have been made by Robertson (1966) (Figure 4.14).   Some typical convection patterns resulting from iso- and contra-rotating crystals and crucibles are shown in Figure 4.15 (Carruthers and Nassau, 1968).   From these it is seen the most extensive Taylor-Proudman columns are generated when there is a large differential velocity between crucible and crystal, with the outer region of melt rotating at the same velocity as the crucible whilst strong flows develop in the central region.   Counter-rotation introduces a third region

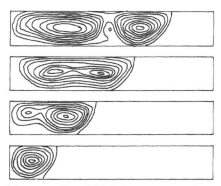

Fig.4.13 Streamlines and solid liquid interface in a 6x1 geometry horizontal Bridgman boat as a function of growth time.   Ra=3x10^4: Pr=0.015: St=140 (after Crochet et al, 1983a).   Reprinted with permission from J.Cryst.Growth published by North Holland, Amsterdam.

immediately below the growing crystal in which the flow pattern is
markedly influenced by crystal rotation rate.

More recently, several workers have used numerical methods to
model the streamline and isotherm distribution in CZ systems
(Langlois, 1981; Kobayashi, 1981; Crochet et al, 1983). Although
there is a fair measure of agreement on the basic patterns observed
for various values of Ra and Pr, detailed modelling of the important
parameters, interface shape and mass transport, depends heavily on
the characteristics of individual crystal-pullers and the procedures
used. A general conclusion, however, is that whilst forced-
convection results in a stable thermal environment, this is achieved
at the expense of faster melt velocities and increased crucible
erosion. In silicon growth, for example, one of the major problems
of CZ growth is the high oxygen content resulting from the
interaction between the melt and the quartz crucible (Hoshikawa et
al, 1981). (In crucible-free float-zoned material the oxygen

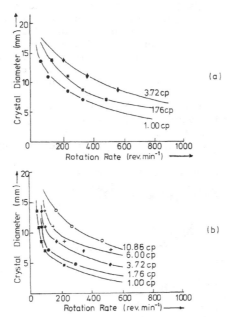

Fig.4.14 Rotation rates required to establish the upward
         flow for crystals of various diameters:
         a) crystal rotation alone at several viscosity values;
         b) crystal and crucible rotation in the same sense at
            several viscosity values (after Robertson, 1966).
         Reprinted with permission from Brit.J.Appl.Phys. published
         by Institute of Physics, Bristol, England.

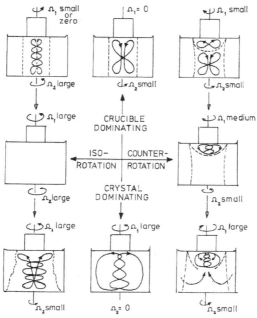

Fig.4.15 Forced-convection flow patterns in CZ melts resulting
         from various combinations of crystal and crucible
         rotation (after Carruthers and Nassau, 1968). Reprinted
         with permission from J.Appl.Phys. published by the
         American Institute of Physics, New York.

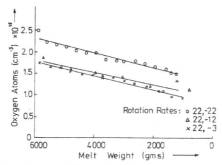

Fig.4.16 Effect of crucible rotation rate on oxygen uptake in
         CZ silicon. Crystal rotation rate 22 rpm (after
         Barraclough, 1982). Reprinted with permission from
         "Aggregation Phenomena of Point Defects in Silicon",
         Eds. E Sirtl and J Goorisen, published by Electro-
         chemical Society, Pennington, New Jersey.

concentration is $<10^{16}cm^{-3}$, but the steep thermal gradients at the
zone edge lead to high dislocation densities). Barraclough (1982)
has demonstrated that the parameter exerting most influence on
oxygen uptake is the crucible rotation rate. The oxygen
concentration rises with increasing rotation rate once a threshold
value of about 12 rpm is exceeded (Figure 4.16). Similar effects
have been observed by Zulehner (1983). Figure 4.17 shows the oxygen
concentration as a function of axial position in the crystal for
several different pulling conditions. The broken curves 1 and 2
show typical distributions in conventional CZ growth whilst curves
4-11 demonstrate improvements due to improved furnace design. The
differences in oxygen concentration of nearly two orders of
magnitude in curves 4-11 are attributable solely to changes in the
crystal and crucible rotation rates. Changes in these parameters
also have a marked effect on the radial and axial distribution of
dopants as illustrated in Figure 4.18 for phosphorus-doped silicon.

Series and Barraclough (1982, 1983) have also examined the
effects of carbon uptake from the carbon heater and crucible support
components (via the gas ambient) and concluded that much of the
incorporated carbon (0.5-3.0×10^{16} cm^{-3}) arose from contamination
of the crucible, charge and melt prior to actual growth. By careful
control of charging and melt-down procedures the carbon content of

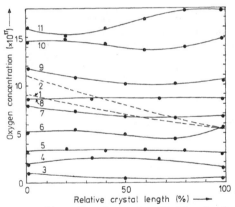

Fig.4.17 Axial oxygen distributions in CZ-Si single-crystals
        grown using a silica crucible. Curves 1 and 2 represent
        typical axial distributions for conventional pulling
        conditions. Curves 3-11 are produced by an improved
        pulling technique developed by Wacker-Chemitronic.
        The pulling of crystals 4-11 differs only in crystal and
        crucible rotation rates. Crystal 3 uses an increased
        Ar pressure (after Zulehner, 1983). Reprinted with
        permission from J.Cryst.Growth published by North
        Holland, Amsterdam.

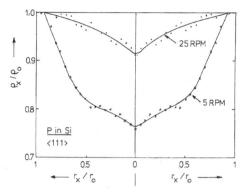

Fig.4.18 Influence of crystal rotation rate on radial resistivity
          variations in two CZ crystals. The pulling conditions
          differ only in rotation rate (after Zulehner, 1983).
          Reprinted with permission from J.Cryst.Growth published
          by North Holland, Amsterdam.

CZ silicon can be reduced to $< 3 \times 10^{15} cm^{-3}$. In an interesting
study of the precipitation characteristics of oxygen during
post-growth annealing, Wilkes (1983) noted that precipitation and
denuded zone formation were retarded in low-carbon material,
suggesting that carbon plays an important role by acting as a
nucleation site.

A further common contaminant in CZ silicon is Fe (Schmidt and
Pearse, 1981), resulting from vapour transport from the hot metal
parts of the puller and condensation on the growing crystal.
Typical concentrations are around $5 \times 10^{15} cm^{-3}$ but can be reduced by
careful puller design.

An alternative method of suppressing convective flow in CZ
melts is to use magnetic damping. The effects of magnetic fields on
melt convection have been studied by Kakutani (1962), Chedzey and
Hurle (1966), Jakeman and Hurle (1972), Wheeler (1984) and Langlois
(1984), and have been applied to the CZ growth of silicon by
Mihelicic and Wingerath (1985), Hurle and Series (1985) and Series
et al (1985). Although fields of >500 Gauss are sufficient to
reduce melt convection very substantially, a current problem is that
the reduced heat transport through the melt necessitates a higher
crucible temperature, leading to increased oxygen uptake from the
quartz.

The use of the Czochralski technique in the growth of volatile
III-V compounds, such as GaAs and InP, was made possible by the
development of the Liquid Encapsulated Czochralski (LEC) process
(Mullin et al, 1970) in which the As or P dissociation loss is

eliminated by encapsulating the melt in B_2O_3 and pressurising the entire growth environment with an inert gas. One of the major problems encountered in LEC growth of these compounds is the high dislocation density (5×10^4-10^5 cm^{-3}) resulting from the high thermal strains produced by the large temperature gradients above the melt. Recently such gradients have been substantially reduced by increasing the depth of B_2O_3 and increasing the heat input to the encapsulant in order to flatten the temperature gradient (Shimada et al, 1984). Axial temperature gradients of 10-25°C cm^{-1} were used and dislocation densities of 1-5×10^4 cm^{-3} were achieved. The use of lower axial gradients implies, however, that the radial gradients are also reduced, increasing the difficulty of diameter control. A further approach to dislocation reduction, which may be used in conjunction with the reduction of temperature gradients, is the addition of an alloying addition to cause lattice hardening. The addition of In to GaAs (Jacob et al, 1981) or Zn to CdTe (Bell and Sen, 1985) have both led to substantial reductions in dislocation content.

CHAPTER 5

VAPOUR PHASE EPITAXY

5.1 Introduction

There is considerable variety and confusion in the nomenclature
of vapour phase processes which reflects the differing
characteristics of the laboratories of origin. For example, in
silicon processing, where the product of vapour phase reactions may
be polysilicon, epitaxial silicon or amorphous oxides or nitrides,
the term CVD (Chemical Vapour Deposition) is applied to all
processes, irrespective of the crystalline or amorphous state of the
product. Where the development of these processes has led to the
introduction of metalorganic source materials the process is known
as MOCVD. In laboratories concerned with single-crystal growth of a
range of materials by vapour phase epitaxy (VPE) the term MOVPE is
more common, emphasising the importance of the epitaxial aspects.
Chemists draw an even finer distinction between OMVPE and MOVPE
since the term "organometallic" refers to those compounds in which
the carbon atom is directly bound to the metal atom, whereas
"metalorganic" is a general term referring to any compound
containing metal atoms in combination with organic radicals.

For the sake of clarity in this chapter, a further generic term
- CVE (Chemical Vapour Epitaxy) - is introduced as a 'first'
subdivision of VPE processes (Table 5.1). This permits the
treatment of silicon processing by pyrolysis and reduction of
trichlorosilane etc under the same heading as MOVPE processes for
III-V and II-VIs, emphasising their common fundamental principles.

Table 5.1   Hierarchy of vapour-phase deposition processes

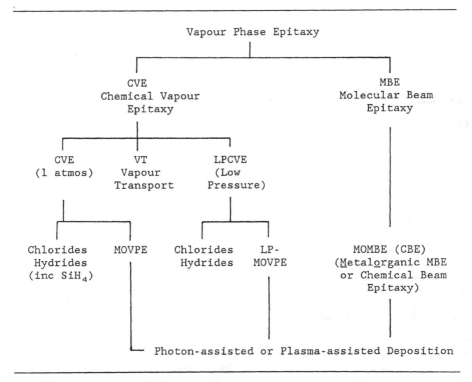

        The essential stages of any vapour phase epitaxy (VPE) process
are:

1)   Generation of precursor atoms or molecules
2)   Transport from source to substrate
3)   Adsorption on substrate surface
4)   Nucleation and growth of epilayer
5)   Removal of unwanted reaction products

    In the simplest VPE systems volatile materials are evaporated
in the high-temperature source region of a closed-tube and
transported down the thermal gradient to deposit on a suitable
substrate located in the cooler region (Figure 5.1a). For growth
from materials of lower volatility, the source materials are

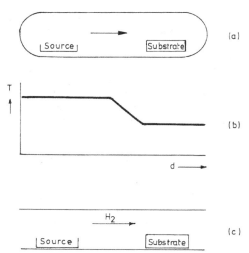

Fig. 5.1   a) Closed-tube vapour transport; b) Idealised form of
source/substrate temperature profile; c) Open vapour
transport system using H_2 carrier gas.

frequently chemically combined with a reactive gas, eg chlorine or
iodine, to ensure efficient vapour transport. The process is
therefore referred to as Chemical Vapour Transport (CVT) or Chemical
Vapour Deposition (CVD). For reversible reactions it is sometimes
performed in closed tubes but it is more frequently used in open
tube systems (Figure 5.1b), in which the precursor molecules are
transported in a H_2 or He carrier gas. Chloride processes have
historically played an important part in the development of
commercial VPE processes, the best-known reactions being

$$2SiHCl_3 \rightarrow 2Si + H_2 + 3Cl_2 \quad (1100/1150°C) \qquad (5.1)$$

$$InCl_3 + PH_3 \rightarrow InP + 3HCl \qquad (650/750°C) \qquad (5.2)$$

$$GaCl_3 + AsH_3 \rightarrow GaAs + 3HCl \qquad (750/900°C) \qquad (5.3)$$

High-temperature vapour-phase processes have been comprehensively
reviewed in the literature. Important works include those by Shaw
(1974) and Kaldis (1974).

Although many commercially successful growth systems have been
developed, the intrinsic problems of these processes lie in the
high source and substrate temperatures required for efficient
vapour generation, transport and growth. In addition to the high
possibility of impurity pick-up from boats and reaction tube walls
resulting from the combination of high temperatures, and the

reactive nature of gases and products (frequently HCl), convection
in the gas stream can lead to variations in the uptake and
transport of reactants and to non-uniform and uncontrollable
growth. The high temperature at the substrate can also result in
graded interfaces (through interdiffusion) and the out-diffusion of
impurities or dopants from the substrate to the layer. Cross-slip
and climb of interfacial dislocations may also occur, giving broad
damage regions and a generally increased dislocation density, whilst
strain induced during cooling from the growth temperature can lead
to further dislocation generation.

It has long been recognised that more reproducible deposition
could be achieved, in principle at least, if gaseous source
materials were used, particularly if laminar flow and lower
temperature deposition conditions could be achieved. In silicon
growth, such techniques developed logically in the late 1950s and
early 1960s from the Siemens process (Hoerni, 1960) already used for
the production of bulk polysilicon according to reaction (5.1).

In the late 1960s and early 1970s Manasevit (1968; 1973) and
Bass (1975) pioneered the use of organic alkyls as sources for the
growth of III-V and II-VI compounds. This led to the development of
modern MOVPE. The advantage of these precursors is that they can be
readily pyrolysed at lower temperatures:

$$(C_2H_5)_3In + PH_3 \rightarrow InP + 3C_2H_6 \qquad (500\text{-}600°C) \qquad (5.4)$$

$$(CH_3)_3Ga + AsH_3 \rightarrow GaAs + 3CH_4 \qquad (500\text{-}700°C) \qquad (5.5)$$

A second advantage is the improved vapour-pressure and flow control
obtained using gaseous sources.

Low deposition temperature is particularly important in the
growth of II-VI compounds, as has been shown by Mullin, Irvine and
co-workers (1981a, 1981b) for the narrow-gap II-VI materials (eg
CdTe, HgTe, $Cd_xHg_{1-x}Te$) and by Wright, Cockayne et al (1982; 1984)
for wide-gap materials such as ZnS and ZnSe. The reactions involved
are:

$$x(CH_3)_2Cd_{(g)} + (C_2H_5)_2Te_{(g)} + (1-x)Hg_{(g)}$$

$$\rightarrow Cd_xHg_{1-x}Te_{(s)} + \text{hydrocarbons (410°C)} \quad (5.6)$$

$$(CH_3)_2Zn_{(g)} + H_2S_{(g)} \rightarrow ZnS + 2CH_4 \qquad (350°C) \qquad (5.7)$$

$$(CH_3)_2Zn_{(g)} + H_2Se_{(g)} \rightarrow ZnSe + 2CH_4 \qquad (350\text{-}550°C) \qquad (5.8)$$

An alternative route to low-temperature deposition involves the
irradiation of a suitable substrate with molecular beams of the
precursors in an ultra-high vacuum (UHV) reaction vessel. The
technique is known as Molecular Beam Epitaxy (MBE). Although some

preliminary experiments were performed as early as 1958, the technique was of little practical significance until the development of reliable UHV equipment, with base pressures of < 10^{-9}Torr, in the mid-1960s. It now finds wide use in the growth of silicon, III-V and II-VI compounds. Its main advantages are the precisely controllable growth rate, making possible the reproducible growth of thin layers to within atomic layer precision, combined with the lack of interdiffusion at the low growth temperatures. These features make it an ideal process for the growth of superlattice structures.

An important side-effect of UHV growth is that it permits *in situ* examination of the substrate surface and epilayer by techniques such as Auger Electron Spectroscopy (AES) (Ploog and Fischer, 1977), Modulated-beam Mass Spectrometry MBMS (Neave and Joyce, 1978) and Reflection High Energy Electron Diffraction (RHEED) (Cho, 1971). In this way MBE has contributed immensely to modern crystal growth science through studies of surface impurities, adsorption and reconstruction. Some of the results of these studies have been discussed in Chapter 2 and their impact on modern growth technology will be briefly explored in Chapter 6.

As might be expected, modern developments in CVE and MBE are aimed at incorporating the advantages of each into the same growth system. To this end, some groups (eg Tsang et al, 1985; Putz et al, 1985 and Kawaguchi et al, 1985) are developing systems using metalorganic compounds as MBE source materials (Metalorganic Molecular Beam Epitaxy, MOMBE) whilst others (Hottier and Theeten, 1980; Robbins et al, 1987) are developing *in situ* surface monitoring techniques such as RAIRS, light scattering and ellipsometry to study surface properties in atmospheric or low-pressure CVE systems.

5.2   Chemical Vapour Epitaxy (CVE)

The general layout of a typical modern CVE system is shown schematically in Figure 5.2. The precursors, many of which are

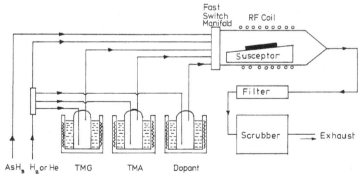

Fig. 5.2   Schematic diagram of an MOVPE reactor for the growth of GaAs and $Al_xGa_{1-x}As$.

highly inflammable and toxic, are contained in thermostatically
controlled baths and fed through mass-flow controlled metering
systems via stainless steel pipework to the reaction vessel. The
substrate is mounted on a carbon susceptor, frequently inclined at a
slight angle to the gas flow. The substrate temperature is
controlled by RF or resistance heating. Both "hot-wall" and
"cold-wall" reactors are used. In the former the reactor wall is
heated, either by radiation from the susceptor or directly by
resistive heating, in order to prevent premature condensation of one
or more of the precursor species. This is necessary in conventional
vapour transport systems to ensure that condensation occurs on the
substrate rather than the reactor walls. Wall-heating is also
required in the growth of $Cd_xHg_{1-x}Te$ to prevent condensation of free
Hg. However, in this case care must be taken to ensure that the
wall temperature is not high enough to promote premature reaction.
Such reactors should perhaps be referred to as 'warm-wall'
reactors. In cases where condensation of the precursors is not a
problem, eg silicon deposition from silane, the reactor wall is
usually water-cooled to prevent premature reaction.

Multilayer growth and dopant incorporation is facilitated by
computer-controlled switching of gas flows. The computer also
provides a safety interlock and monitoring function. Sharply-
profiled epilayer combinations (eg superlattices, laser confinement
structures etc) cannot be produced by rapid switching alone, since
the memory effect associated with residual gas in the downstream
side of the switches leads to composition grading. Careful reactor
and gas-inlet manifold design are therefore required to ensure that
switches are positioned as close to the reaction vessel as possible
and that the minimum of premixing occurs (Griffiths et al, 1983).
Unfortunately, uniformity in both thickness and alloy composition
over large areas is only achievable with thorough pre-mixing of the
precursors and a stable flow regime. Reconciling these two opposing
requirements is one of the challenges facing the development of
large area CVE growth. One way of minimising the effect of gas flow
variations is to ensure that growth occurs in the kinetically-
limited regime (Chapter 2), but the restriction to low growth rates
is not always acceptable in practical growth systems, nor, in other
cases, easily achieved.

5.2.1 Reactor design

Much time and effort has been deployed in the design of reactor
cells and the analysis of their fluid flow and material deposition
characteristics. Effective design is based on a combination of
minimum gas 'dead space', uniform temperature distribution over the
substrate area and laminar gas flow. Several technological routes
have been used to obtain uniform substrate temperatures (see for
example Beneking et al, 1981) but their success or otherwise is
empirically determined and will not be discussed further here. The

four basic types are illustrated in Figure 5.3. Simple cells such
as the horizontal or vertical reactors are used for R&D purposes,
where the number of epilayers required is small or where alloy
growth demands very reproducible gas transport conditions. The
pancake and barrel reactors are favoured for high-volume production
runs, particularly in silicon deposition. In general the gas flows
in these reactors are less well characterised and growth rates are
more variable.

5.2.1.1 The horizontal reactor

This type of reactor is the simplest and therefore most
completely characterised in terms of the gas velocity, temperature
and reactant mass distributions. The fluid flow pattern results
from interaction between the horizontal flow of the input gas stream
and the vertical buoyancy-driven convection component due to the

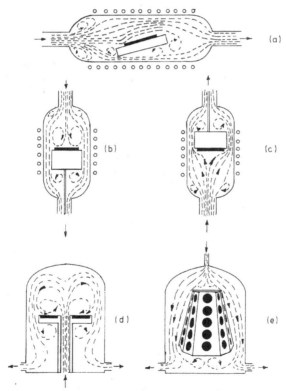

Fig. 5.3 Basic CVE reactor types a) horizontal, b) top-fed vertical
c) bottom-fed vertical, d) pancake, e) barrel.

presence of the heated substrate in the lower part of the growth tube.

The onset of turbulence in horizontal tube flow-patterns in the absence of any buoyancy-driven convection occurs at Re > 2000 where Re is the Reynolds number. For CVE flow rates, typically 1-50 cm sec^{-1}, Re_{H_2} and Re_{He} are 30-50 whilst for Ar and N_2 higher values of about 100 are observed, but all lie well within the laminar flow regime. In the central region of the susceptor the temperature gradient is normal to the tube axis and the buoyancy-driven convection is characterised by the Rayleigh number Ra. Again, the critical value for convection, Ra_c > 1700, is much higher than the values obtained in typical horizontal reactors. (Since Ra α ΔT, the buoyancy-driven convection is stronger in cold wall reactors than in hot-wall reactors). Nevertheless, the combination of the two driving forces results in quite marked convection patterns which have been studied by several groups (see, for example, Sparrow et al, 1959, Everstejn et al, 1970, Berkman et al, 1977, Ban, 1978).

Using thermocouples and mass spectrometric probes, combined with stream-visualisation techniques based on TiO_2 "smoke", Ban (1978) showed that a steep thermal gradient exists in the vicinity of the susceptor surface (Figure 5.4a) and that at high Gr/Re^2 ratios a spiral flow pattern develops, as a result of the increasing dominance of buoyancy-driven convection (characterised by increasing Grashof number Gr) (Figure 5.4b). The use of Gr/Re^2 to characterise

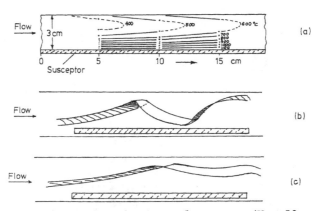

Fig. 5.4 a) Isotherms in a horizontal reactor (V = 50 cm sec^{-1}: T_S = 1200°C), b) TiO_2 smoke streamlines for T $\leqslant$ 800°C; Gr/Re^2 = 0.47, 0.56, 4.14, 12.5, c) TiO_2 streamlines for T $\geqslant$ 800°C; Gr/Re^2 = 0.16, 0.21, 0.53 (after Ban, 1978). Reprinted from J.Electrochem.Soc. by permission of the publisher, The Electrochemical Society.

interactive convective flows was proposed by Sparrow et al (1959) who defined three regimes:

$$0 \ < Gr/Re^2 < 0.5 \qquad \text{Forced-convection dominant}$$
$$0.5 < Gr/Re^2 < 16 \qquad \text{Mixed convection}$$
$$Gr/Re^2 > 16 \qquad \text{Buoyancy-driven convection}$$

The use of this analysis for horizontal systems as in Ban's work has been criticised by Westphal (1983) on the grounds that it was originally derived for flow in the vicinity of a vertical wall. Although the interpretation in terms of Gr/Re^2 is therefore suspect, similar flow patterns have been observed by other workers including Giling (1982) who used more sophisticated laser interferometry techniques to study the gas flow in real time (Figure 5.5). One of the most striking features of Giling's analysis was the marked

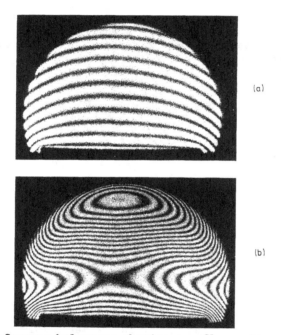

Fig. 5.5   Interference holograms showing gas flow patterns and temp-
           erature profiles in a horizontal reactor.   a) H_2(or He) at
           low flow velocity v = 10 cm sec^{-1}. Air-cooled reactor with
           $T_{susceptor}$ = 1350K, susceptor length = 20 cm.   b) N_2(or Ar),
           low flow velocity, v $\leqslant$ 4 cm sec^{-1}. Air-cooled reactor with
           $T_{susceptor}$ = 1300K susceptor length = 20 cm (after Giling,
           1982).   Reprinted from J.Electrochem.Soc. by permission of
           the publisher, The Electrochemical Society.

difference in flow characteristics when H_2 or He are replaced by Ar
or N_2 as the carrier gas. The N_2 and Ar flows are highly unstable
at flow rates above about 4 cm sec^{-1}. Even at lower velocities
stable convection cells can be observed.

An important consideration in epitaxial reactor design is the
'entrance length' for the development of the equilibrium flow
profile upon entry to the reactor, or following perturbation by the
leading edge of the susceptor or any other obstruction in the flow
tube. According to Schlichting (1968), flow in a channel of height
h develops to its equilibrium velocity profile at distance x_v given
by

$$x_v = 0.04h.Re \qquad (5.9)$$

for typical reactor height h = 2.5 cm and v = 10-40 cm sec^{-1}.
Giling and other workers derive entrance lengths ranging from 0.2 cm
(He: T = 1400K : v = 10 cm sec $^{-1}$) to 70 cm (Ar:T = 300K: v = 40 cm
sec^{-1}). The distance over which the equilibrium *temperature* profile
x_T develops has been shown by Hwang and Cheng (1973) to be given by
the expression:

$$x_T = 0.28h.Re \qquad (5.10)$$

ie seven times greater than x_v. The corresponding entrance length
for mass transfer, x_m is approximately $2x_T$ and Chen (1984) has
computed a growth rate $\dot{N}$ given by

$$\dot{N} = c_1 (1-c_2) \frac{(Gz.Sc)}{Pr}^{c_2} \frac{D}{h} \cdot \frac{P_o}{RT_m} \qquad (5.11)$$

for $0 \leqslant x \leqslant x_m$ (where $c_1 = 1.185$ and $c_2 = 0.445$) and

$$\dot{N} = \frac{4.86D}{h} \cdot \frac{P}{RT_m} \qquad (5.12)$$

for $x > x_m$

In the above equations: Gz is the Graetz number, Sc is the
Schmidt number, D is the diffusion coefficient of the chemical
species in the carrier gas and T_m is the mean fluid temperature.
Despite the use of averaged temperatures and gas properties, the
calculated curve shows excellent agreement with Eversteyn et al's
(1970) experimental results, both for the case of a horizontal
susceptor and (with slight modifications to equations 5.11 and 5.12)
to growth on inclined susceptors. These are frequently used to
create increased downstream flow velocities to counter the effects
of chemical depletion in the gas stream and achieve improved growth
uniformity (Figure 5.6).

Hitchman (1980) has shown that the reason such good agreement
can be achieved using 'averaged' properties and temperatures in real
situations involving steep thermal gradients is that the
relationship between the averaged mass flux j_{av} and that calculated
using variable gas properties and temperature, $j_{variable}$, is of the
form

$$j_{av} = \alpha^{1/2} \, j_{variable} \qquad\qquad (5.13)$$

where $\alpha = \mu\rho/\mu_\infty\rho_\infty$. In this, μ and ρ are the viscosity and density
at a position in the stream near the susceptor, whilst μ_∞ and ρ_∞ are
the corresponding values at a point in the bulk fluid. Since μ and
ρ show opposing trends with temperature, $\alpha^{1/2} \sim 0.8\text{-}1.0$ over a wide
temperature range.

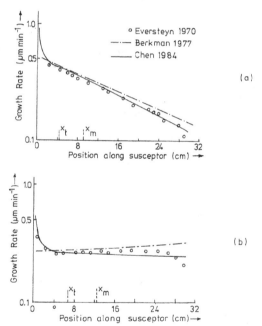

Fig. 5.6   Growth rate of epitaxial silicon as a function of position
on a) horizontal and b) inclined susceptor; x_t, x_m are temperature and mass entrance lengths (after Chen, 1984).
Reprinted with permission from J.Cryst.Growth, published
by North Holland Publishing Co., Amsterdam.

5.2.1.2 The vertical reactor

The simplest form of vertical reactor is the chimney reactor, which is a bottom-fed reactor (ie gas input below substrate). Several reviews of the characteristics of this type of reactor are available (see, for example, Sugawara, 1972; Wahl, 1977; Wahl and Hoffmann, 1980; Hitchman and Curtis, 1981 and 1982). Better control of gas flow can be achieved in top-fed reactors since the downward, forced-convective flow can be adjusted to balance the upward buoyancy-driven convection. Unfortunately, although the flow-pattern is improved, this geometry is not acceptable for systems prone to premature gas-phase reaction, since any solid particles generated fall onto the substrate giving spurious nucleation and poor quality growth.

5.2.1.3 The pancake reactor

As is evident from Figure 5.3, the gas flows in this reactor are extremely difficult to quantify and show a marked dependence on individual reactor geometries and gas flow rates. Indeed, the aim of the reactor designer is to achieve thorough gas mixing by injecting the gas through a central orifice in such a way that it impinges on the bell-jar surface and is swept down the sides and across the substrate, the circulation being repeated several times before exhaustion. Several substrates can be accommodated on the horizontal carousel, making the technique important for high through-put production runs. Although suitable for silicon pyrolysis, it is less satisfactory for compound semiconductor growth since the large wall-area/gas interaction can lead to premature reaction and non-uniform deposition rates. (The typical growth profile of a pancake reactor shows a much greater rate of deposition at the centre than at the edge).

Analytical work has concentrated mainly on the effect of gas flow, temperature gradient and substrate preparation on autodoping the out-diffusion or vapour transport of impurities or dopants from the substrate to the layer (Bozler, 1975; Chang, 1985).

5.2.1.4 The barrel reactor

Despite its apparently different geometry, the gas dynamics of the barrel reactor are in fact very similar to those of a horizontal reactor with inclined substrate. Provided that the space between the substrate carousel and the reactor wall is small, the flow is essentially laminar over the whole range of growth conditions used and extremely good uniformity of growth and composition can be achieved. Analyses of gas flows in these reactors have been given by Fuji et al (1972); van der Putte et al (1975); Manke and Donaghey (1977) and Juza and Cermak (1982). Powerful computational methods are now available for the calculation of mass transport and deposition rates in CVE reactors. In many cases there is good agreement between the calculated and observed growth rates (Figure 5.7) or calculated and observed doping profiles (Reif and Dutton, 1981).

The general design rules evolving from the above analyses can be summarised as follows:

a. Whenever possible, growth should be performed in the kinetically-limited regime, so that the rate is independent of flow conditions.

b. In the mass-flow or diffusion-limited regimes reactors should be designed to ensure minimum free height above the susceptor, which should be positioned at a distance $x > x_m$ from the reactor entrance.

c. Similarly, substrates should be positioned at a distance $x > x_m$ from the leading edge of the susceptor.

d. Leading edges of susceptors, substrates etc should be aerodynamically contoured to minimise disruption of the gas flow.

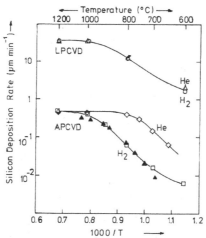

Fig. 5.7   Silicon deposition rate as a function of temperature, total pressure and carrier gas. Open symbols are the results obtained from Coltrin's theoretical model. Solid triangles are experimental data from van der Brekel (1978). Solid curves are drawn to distinguish between low-pressure CVD results (LPCVD) and atmospheric pressure CVD (APCVD) and between H_2 and He carrier gases (after Coltrin 1984). Reprinted from J.Electrochem.Soc. by permission of the publisher, The Electrochemical Society.

e. To maximise the response speed in gas-switching for multilayer growth, reactor dead-space should be minimised and switching valves placed as close to the reactor inlet as practicable.

5.2.2 Low-pressure chemical vapour epitaxy (LPCVE)

Over the last decade several workers have reported improved growth of silicon (Townsend and Uddin, 1973) and III-V compounds Hersee and Duchemin, 1982; Hersee et al, 1983; Razeghi et al, 1983) from the use of reduced pressures (1-100 Torr). These improvements in thickness and compositional uniformity result from a variety of causes, including reduced Rayleigh number and elimination of mass transport and boundary layer effects.

Curtis and Dismukes (1972) have shown that

$$Ra \propto p^2 T^{-m} \tag{5.13}$$

where p is the total gas pressure, T the stream temperature and m is a constant such that $4.3 < m < 4.8$. Pressure reduction is, therefore, an effective way of reducing buoyancy-driven convection characteristic of high Ra.

To a first approximation, the growth is mass-transport controlled for $k\delta/D \gg 1$ and kinetically-controlled for $k\delta/D \ll 1$ where k is the thermal conductivity, δ is the boundary layer thickness and D is the diffusion coefficient of the precursor species in the carrier gas. By operating at low pressure the diffusion coefficient can be increased by a factor of three or more, transferring growth control to the kinetically limited regime.

The combined result of these effects is that the growth temperature can be substantially reduced, with consequent reduction in autodoping and interface dislocation multiplication. For example, SiH_4 pyrolysis, typically carried out at $T_g > 1050°C$ at atmospheric pressure, yields good single crystal material at temperatures as low as 775°C (Donahue et al, 1984), when the system pressure is reduced to about 10 Torr.

In Donahue's work it was found that the achievement of good crystallinity in LPCVE was at least partly due to improved substrate surface preparation (deoxidation), in this case by plasma etching. However, Nagao et al (1985) have shown that enhanced oxide removal can be achieved using only *thermal* pretreatment at low pressure. Good crystallinity was obtained in films grown at 930°C (p=40-760 Torr) following substrate deoxidation in a 10 minute prebake at 900°C with a system pressure of 40 Torr.

A further advantage of LPCVE is found in improved desorption. In any vapour-phase reaction, the removal of the unwanted reaction products from the growth surface presents a potential problem, eg H_2 in silane pyrolysis, C or hydrocarbons in the decomposition of metalorganics. In fact, the first use of LPCVE was to remove adsorbed hydrogen during silane pyrolysis, permitting improved growth at lower temperatures   (Townsend and Uddin, 1983).

Whether or not the use of LPCVE is preferable to atmospheric pressure (or even higher pressures) depends on the role played by the boundary layer. More specifically, Lee (1984) has suggested that LPCVE is only beneficial in improving uniformity of deposition when the growth rate is diffusion limited and $G \propto p^n$ where $n > 0.5$, p being the total pressure in the growth system.

5.2.3 Plasma-enhanced chemical vapour epitaxy (PCVE)

Lower temperature deposition with improved uniformity can sometimes be achieved using an RF discharge to generate a plasma in the immediate vicinity of the growth surface. A suitable modification of the basic CVE reaction cell is shown in Figure 5.8.

The gas phase reactions stimulated by the plasma are extremely complex and difficult to monitor. The effects of the impingement of charged, energetic particles on the growth surface in terms of adsorption, implantation, surface mobility and surface reconstruction are also complex. For these reasons the technique has found widespread application in LPCVD systems for the deposition

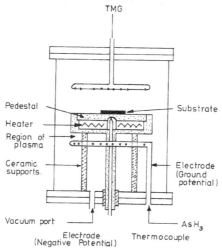

Fig. 5.8   Plasma-enhanced CVE reactor for growth of GaAs from TMG and AsH_3 (after Pande and Seabaugh, 1984). Reprinted from J.Electrochem.Soc. by permission of the publisher, The Electrochemical Society.

of polycrystalline or amorphous oxides or nitrides (Matsuo and
Kiuchi, 1983; Wilkes 1984) but is less well-developed for CVE growth.

5.2.4 Vapour-phase reactions in MOVPE growth

In the previous section we concentrated upon the problems of
thickness and growth-rate uniformity resulting from failure to
control the gas flow patterns in the reaction cell. Other potential
problems of MOVPE growth are carbon contamination, polymer
production and premature reaction.

In the growth of GaAs from the reaction

$$(CH_3)_3Ga + AsH_3 \rightarrow GaAs + 3CH_4 \qquad (5.14)$$

the incorporation of C as a by-product of the hydrocarbon
reaction-product decomposition can result in p-type carrier
concentrations $>10^{17} cm^{-3}$. Ways of reducing this effect are
discussed further in Chapter 6. When MOVPE growth was applied to In
compounds, eg:

$$(CH_3)_3In + PH_3 \rightarrow InP + \text{hydrocarbons} \qquad (5.15)$$

Manesevit and Simpson (1973), Cooper et al (1980) and other workers
found that the reaction products were long-chain polymers which
deposited as an adhesive film on the substrate surface and reaction
vessel walls. Moss (1983) demonstrated that this problem could be
circumvented by the selection of a different phosphorus precursor,
eg $(C_2H_5)_3P$, such that the reaction produced an intermediate Lewis
acid-base adduct, which decomposed to give InP without polymer
formation, viz:

$$(CH_3)_3In + (C_2H_5)_3P \rightarrow (CH_3)_3In.(C_2H_5)_3P \qquad (5.16)$$
$$\underline{\qquad\qquad\qquad\qquad}$$
$$\text{adduct}$$

$$(CH_3)_3In.(C_2H_5)_3P \rightarrow InP + 3CH_4 + 3C_2H_4 \qquad (5.17)$$

Adduct formation is playing an increasingly important role in MOVPE
growth since such compounds can be easily synthesised and are
generally less toxic and more stable than their alkyl precursors.

Adduct formation may also explain some of the observations made
by Irvine et al (1981, 1983) in the growth of CdTe and HgTe. These
authors noted that the efficient pyrolysis of $(C_2H_5)_2Te$ required
temperatures in excess of 410°C, but in the presence of $(CH_3)_2Cd$ the
required temperature was reduced to 350°C, suggesting the formation
of an adduct $(CH_3)_2Cd.(C_2H_5)_2Te$. This permitted growth of CdTe at

$350°C$ whereas HgTe growth required breakdown of $(C_2H_5)_2Te$ and hence temperatures $> 410°C$. In the growth of $Cd_xHg_{1-x}Te$ the formation of $(CH_3)_2Cd.(C_2H_5)_2Te$ (or adducts of other tellurium metalorganics) having low pyrolysis temperatures may be a disadvantage since the reactor wall must be heated to $> 250°C$ to prevent Hg condensation.

The choice of different precursors can lead to different reactions and reaction products. Carrier gases may also participate in the reaction process. For example, Yoshida et al (1985) have shown that the decomposition of $(CH_3)_3Ga$ and $(C_2H_5)_3Ga$ in the H_2 and N_2 occurs by different routes.

Decomposition of $(CH_3)_3Ga$ in hydrogen occurs by hydrogenation according to the reaction

$$CH_3{\atop{CH_3}}{\atop{CH_3}}\!\!\!\raise2pt\hbox{$>$}Ga + H_2 \xrightarrow[450°C]{} \begin{array}{c}CH_3H\\CH_3H\\CH_3\end{array}\!\!\!\raise2pt\hbox{$>$}Ga \rightarrow CH_3Ga + 2CH_4 \qquad (5.18)$$

giving methane as the primary reaction product. In N_2, however, decomposition is purely dissociative:

$$CH_3{\atop{CH_3}}{\atop{CH_3}}\!\!\!\raise2pt\hbox{$>$}Ga + N_2 \xrightarrow[550°C]{} \begin{array}{c}CH_3\\CH_3\\CH_3\end{array}\!\!\!\raise2pt\hbox{$>$}Ga \rightarrow CH_3Ga + C_2H_6 + N_2 \qquad (5.19)$$

giving ethane C_2H_6 as the primary product. The required temperature for complete decomposition is also higher. By contrast $(C_2H_5)_3Ga$ decomposes by β elimination in both carrier gas species, resulting in the formation of ethylene, C_2H_4:

$$C_2H_5{\atop{C_2H_5}}{\atop{C_2H_5}}\!\!\!\raise2pt\hbox{$>$}Ga \xrightarrow[300°C]{} \begin{array}{c}C_2H_5\\CH_3\;CH_2\\CH_3\;CH_2\end{array}\!\!\!\raise2pt\hbox{$>$}Ga \rightarrow C_2H_5Ga + (CH_3)_2Ga + C_2H_4$$
$$\qquad (5.20)$$

One of the inherent problems of residual gas analysis is that the analyser is remote from the reaction zone. The ion source of the mass analyser may also promote further dissociative reactions. For these reasons there is much emphasis on the development of diagnostic probes such as laser-induced fluorescence which can give information on the species present in the reaction zone. Such techniques will be discussed further in the next chapter. Nevertheless, residual gas analysis has proved extremely useful in

identifying the different reactions and reaction products occurring
in different temperature regions and with different precursors and
in helping to select the optimum deposition reaction for good
crystal growth.

5.3 Molecular Beam Epitaxy

Modern MBE growth systems (Figure 5.9) contain three high-
vacuum compartments. The first is a load-lock chamber operated at
around 10^{-6} Torr for loading specimens with minimum disturbance of
the UHV conditions in the other two chambers. The second chamber is
a substrate preparation and assessment chamber operated at $<10^{-10}$
Torr where specimens can be thermally, plasma, or ion-beam etched to
remove surface damage and oxides, and where RHEED and Auger studies
can be performed to establish details of surface reconstruction and
surface impurities. The third chamber, the growth chamber, contains
a rotating, heated, substrate stage, which is positioned to inter-
cept the molecular beams generated in the surrounding Knudsen effusion cells (Figure 5.10) containing the source and dopant species.

The chemical species contained in the molecular beam emitted
from the effusion cell depend on the source material and the cell
temperatures. As we shall see below, modern preference is to use
dimers rather than tetramers for group V and VI materials, eg As, P,
Te. It is therefore becoming common practice to attach a
high-temperature, heavily-baffled "cracker" cell to the forward end

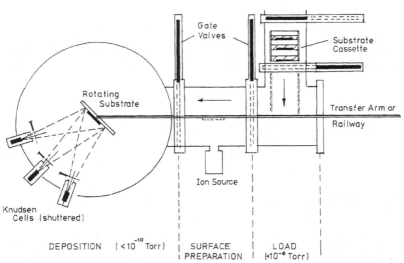

Fig. 5.9   Schematic diagram of a three-chamber Molecular Beam Epitaxy
(MBE) equipment.

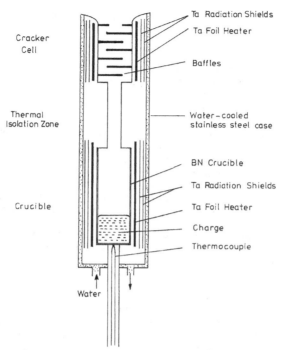

Cracker Cell

Ta Radiation Shields

Ta Foil Heater

Baffles

Thermal Isolation Zone

Water-cooled stainless steel case

BN Crucible

Ta Radiation Shields

Crucible

Ta Foil Heater

Charge

Thermocouple

Water

Fig. 5.10 A Knudsen effusion cell for molecular beam generation, showing the high temperature 'cracker' cell.

of the Knudsen cell to promote thermal dissociation (see for example Neave et al, 1980). In many cases, the conversion efficiency is ≥99.9%.

To obtain thin-layer growth, sometimes only a few Angstroms in thickness, rapid switching of the incident precursor and dopant beams is achieved by computer-controlled, electromagnetically operated, stainless-steel shutters positioned close to the cell orifice. Further details of the construction of MBE equipments may be found in several reviews, including those by Ploog (1980); Chang and Ploog (1985) and Parker (1985). An introductory article by Cho in the last of these summarises some of the important milestones in the development of the technique and its impact on modern crystal growth. A detailed description of the construction and operation of a modern MBE system is given by Davies and Williams (1985). The importance of MBE in the growth of semiconductor materials cannot be overstated. Cho (1985) lists some thirty elements and alloy systems grown by the technique. Complex multilayer device structures, including double-heterostructure lasers, high electron mobility

transistors (HEMTs) and many superlattice structures have been
fabricated in MBE-grown material.

From the crystal-grower's viewpoint, however, the most
significant advances brought about by MBE lie in the ability to
generate, identify and monitor the incident and desorbed fluxes of a
single atomic or molecular species, so that its sticking coefficient
and surface mobility can be studied. The UHV environment eliminates
complex vapour transport variations and permits, through Auger
analysis, a detailed knowledge of the adsorbed species on the
substrate surface. Finally, techniques such as angle-resolved
photo-electron emission spectroscopy (ARPES) and reflection
high-energy electron diffraction (RHEED) can be used to monitor
surface reconstruction and elastic strain during substrate
preparation and subsequent epitaxial growth. For the first time,
experimental conditions can be reliably and reproducibly
established. This enables us to study the differences responsible
for three-dimensional island growth (which usually results in
inferior epitaxy), or in controlled two-dimensional layer growth.
The recent development of lattice-imaging, high-resolution
transmission electron microscopy enables us to examine the
substrate-epilayer interface on a near-atomic scale.

Since the temperatures used in MBE always lie in the
kinetically-limited regime (Figure 2.3), the quality of the
epilayers produced depends critically on the composition and flux of
the molecular beam (which determines the concentration on a species
of the surface adsorbate), the quality of the substrate surface and
the efficient removal of reaction products or excess vapour phase
species.

5.3.1 The molecular beam

The incident flux F from each Knudsen cell is given by the
expression

$$F = \frac{ap}{\pi d^2 \sqrt{2\pi m \ kT}} \ . \ \cos \theta \ [\text{molecules cm}^{-2}\text{s}^{-1}] \qquad (5.21)$$

where a is the area of the cell aperture, p is the equilibrium
vapour pressure of the precursor, d is the distance from orifice to
substrate, m is the mass of the diffusing species, T the cell
temperature (K) and θ the angle of incidence of the beam relative to
the substrate normal. Typical fluxes are of the order of 10^{16}
molecules cm^{-2}s^{-1} for the precursor elements and 10^{11}-10^{12} molecules
cm^{-2}s^{-1} for dopant species. The resultant growth rates are usually
in the range 0.1 - 1.0 μm hr^{-1}. Careful equipment design is
necessary to ensure a uniform arrival rate of the precursor species
and this is usually achieved by a combination of substrate rotation
and Knudsen cell design (with particular emphasis on constant and
reproducible temperature). To further control growth rate and to

minimise contamination, the substrate is usually surrounded by liquid-nitrogen-cooled cryopanels. Again, care must be taken to ensure that the temperature of these remains constant throughout the growth run.

GaAs may be grown from Knudsen sources containing elemental Ga and As, or from a single GaAs source. Whereas the Ga flux is monatomic, the As derived from elemental sources is in the form of As_4 tetramers. In contrast, As from GaAs dissociation is in the form of As_2. However, the dimer flux from GaAs sources contains around 15% of the group III element, the variation in which makes quantitative kinetic studies difficult and in most cases the required As_2 dimers are produced by cracking the As_4 molecules from the elemental source (Neave et al, 1980). It is now well established that different film growth characteristics and properties arise when As_2 is used as the molecular group V species rather than As_4. These differences include:

a) Lower concentrations of the deep levels M_1, M_3 and M_4 in GaAs grown from the As_2 (Neave et al, 1980; Kunzel et al, 1982).

b) Ge, an amphoteric dopant, is incorporated on As sites in As_4-grown material but occupies Ga sites in As_2 grown material, behaving as a donor impurity and leading to a reduced level of autocompensation.

c) Surface reconstruction data from ARPES shows a stronger S_2 peak in material grown from As_2 (Joyce, 1985). Since S_2 is related to emission from an As-derived surface state, this is circumstantial evidence for a higher As concentration in the surface.

d) The As sticking coefficient measured by modulated beam mass spectroscopy (Joyce and Foxon, 1975; Foxon and Joyce, 1977) on a Ga-stabilised surface is always $\leqslant 0.5$ for an As_4 flux, whereas for an As_2 flux it approaches unity at high temperature. This difference has been explained by Joyce et al (1978) in terms of the different adsorption mechanisms illustrated in Figure 5.11. The dissociation of As_2 results in one As atom per Ga site and monolayer coverage ($S_{As_2} = 1$) whereas the pair-wise dissociation of each tetramer requires the presence of two adjacent Ga sites, giving a maximum value of S_{As_4} of 0.5. Effects b)-d) are all consistent with a lower As surface concentration in the case of As_4-grown material.

InP has been produced by MBE using both P_4 (McFee et al, 1977) and P_2 sources (Roberts et al, 1981; Tsang et al, 1982). As in the case of GaAs growth, the dimer appears to stabilise the InP surface to a higher temperature than the tetramer, presumably by more effective surface coverage. Unlike GaAs, an InP source is not a particularly good source of the group V dimer. In the temperature range 530-730°C P_2 is the dominant species but approximately 10% of

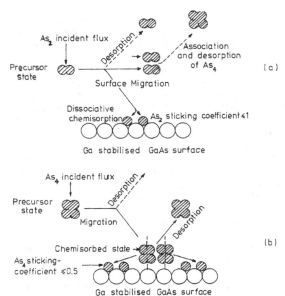

Fig. 5.11 Model of the growth of GaAs from molecular beams of a) As_2 and b) As_4 (after Joyce et al, 1978).

the total phosphorus flux is composed of tetramers. A further potential problem is that above approximately 360°C the evaporation of InP is incongruent leading to rapid depletion of phosphorus at higher temperatures. The vapour pressure of P and As as a function of Knudsen cell temperatures are shown in Figure 5.12. The usual source of P_4 is red phosphorus, but problems can arise in the use of this source since even at 150°C $p(P_4) \simeq 10^{-3}$ Torr, making bake-out of the UHV system difficult. Furthermore, excess P_4 condenses on cryopanels in the form of white phosphorus leading to high vapour pressures on desorption. In contrast, P_2 condenses as red phosphorus producing much lower vapour pressures on subsequent desorption. This fact, together with its growth surface adsorption characteristics, which appear similar to those of As_2 in GaAs growth, make P_2 the preferred source. It can be generated by thermal cracking of P_4 at temperatures >900°C, under which conditions the conversion efficiency is about 95%. Further details of InP MBE are given by Stanley et al (1985).

5.3.2 Substrate preparation

Correct substrate preparation is crucial to the success of epilayer deposition by MBE since the low diffusion rates at the temperatures involved mean that defects once introduced cannot

easily be eliminated. As discussed in Chapter 3, poor mechanical
polishing techniques can lead to surface dislocation densities > 10^9
cm^{-2} and the damage depth can be several microns in soft materials.
Many of the chemical etches or chemo-mechanical polishes used to
remove surface-damage lead to non-stoichiometric or oxidised
surfaces which must be removed prior to growth. In other cases such
as silicon, the clean surface is highly reactive and slices are
therefore supplied with a thin (~50Å) oxide film. Although this can
be removed by heating to > 800°C, Hardeman et al (1985) have shown
that the removal is frequently incomplete on a microscopic scale.
High-temperature heat-cleaning also increases the possibility of
dislocation generation or elastic strains leading to wafer warpage.
Ar-ion sputter-cleaning is frequently employed as an alternative,
lower temperature, technique, but the advantage is somewhat reduced
by the need to anneal at about 850°C to remove ion-damage. More
recently the development of oxidising treatments yielding much
thinner (~5Å) oxides which can be desorbed at 750°C have led to the
reintroduction of thermal cleaning techniques (Shiraki, 1985).

In compound semiconductor substrate preparation, ion
bombardment frequently leads to preferential loss of the higher
vapour-pressure group V or VI element, resulting in excess metal
accumulation on the growth surface. This is particularly so in the
case of InP or InSb substrates, where ion milling leads to
In-droplet formation (Williams et al, 1985), which generates
stacking-faults during subsequent MBE growth (Figure 5.13).

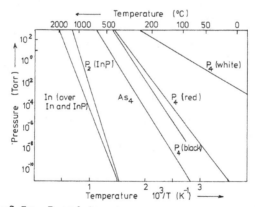

Fig. 5.12 Plot of In, P and As vapour pressures as a function of
reciprocal temperature. The curves labelled P_4 (white), P_4
(red) and P_4 (black) are the equilibrium vapour pressures
of P_4 over white, red and black phosphorus respectively.
The equilibrium vapour pressure of As_4 over elemental As
is shown for comparison (after Stanley et al, 1985).

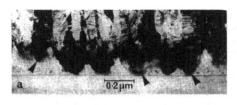

Fig. 5.13 Cross-sectional transmission electron micrographs of CdTe
layers on InSb substrates a) T_S = 150°C, interface region
(arrows indicate Te precipitates at the base of stacking
faults), b) T_S = 180°C showing fault-free growth (after
Williams et al, 1985). Reprinted with permission from
J.Vac.Sci.Technol. published by American Institute of
Physics, New York.

5.3.3 Surface reconstruction and adsorption

The relaxation of the interatomic bonds between atoms in the
surface and near-surface regions of a material leads to a surface
structure of lower symmetry than the bulk. The magnitude of this
relaxation is inversely related to the bond strength of the material
and in soft materials such as GaAs and CdTe may extend to a depth of
a few atomic layers. The relaxation is also dependent upon crystal
orientation and the presence of adsorbed precursor, dopant or
impurity species. In polar materials the composition of the surface
plane (anion- or cation-rich) may also affect the surface
reconstruction. Conventionally, the notation used to describe the
reconstructed surface is to define a surface unit cell whose
primitive vectors are m and n times those of the bulk lattice. The
surface is then said to be an m x n reconstruction (Wood, 1964).
The required information is usually derived from RHEED patterns
taken in two orthogonal directions (Figure 5.14). The electron
wavelength is given by

$$\lambda = \left[\frac{150}{V(1+10^{-6}V)} \right]^{1/2} \qquad [\text{Å}] \qquad (5.22)$$

where V is the accelerating potential. The periodicity "m" (or "n"
in the orthogonal direction) of the surface structure is given by

$$m = \lambda L/t \qquad (5.23)$$

where L is the distance between crystal surface and phosphor screen
and t is the electron diffraction streak spacing.

The most frequently studied and best characterised surfaces are
those of GaAs (Neave and Joyce, 1978; MacRae and Gobeli, 1966;
Larsen et al, 1983; Joyce, 1983 and Joyce, 1985). (001) is a polar
face in GaAs and the surface may be terminated by either a Ga atom
layer or an As atom layer. In practice Ga-surface layers are

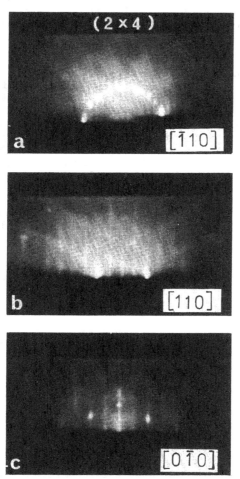

Fig. 5.14 RHEED patterns from a (2x4) reconstructed GaAs surface for
          two orthogonal <110> directions (a and b) and the [010]
          direction (c) (courtesy of C Whitehouse, RSRE)

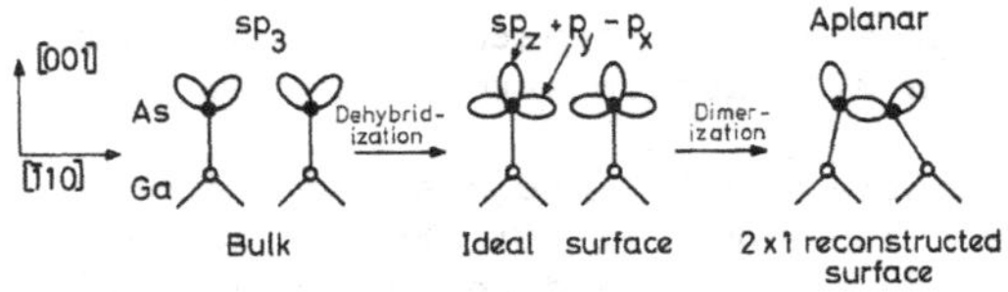

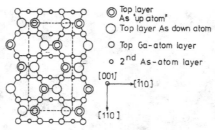

Fig. 5.15 Schematic diagram of the dehybridisation and dimerisation
of the GaAs molecule leading to an aplanar dimer structure
(after Joyce, 1985). Reprinted with permission from
"Molecular Beam Epitaxy and Heterostructures", (Eds.
N Chang and K Ploog) published by Martinus Nijhoff,
PO Box 163, 3300 AD Dordecht, The Netherlands.

unstable, since under As-deficient conditions free-Ga clusters are
formed, whilst in an As-rich environment the Ga readily adsorbs As
to form an As-terminated surface. The most frequently observed
surface reconstruction of the As-stabilised (001) GaAs surface is the
(2x4) structure, ie the surface primitive cell vector [110] is twice
the bulk cell vector for this direction and the [110] vector is four
times the length of the corresponding bulk vector. The surface
bond-tilting responsible for this symmetry reduction is shown
schematically in Figure 5.15. Since only (110) planes are
illustrated the reconstruction is (2x1). A similar (2x1)
reconstruction of (001) silicon has been observed by Schlier and
Farnworth (1959) and is also attributed to dimer formation. An
(001) surface map showing the 001 (2x4) reconstructed arrangement of
surface and second layer atoms is shown in Figure 5.16.

The (001) face shows many different reconstructions varying with
composition (As coverage) and temperature (Drathen et al, 1978;

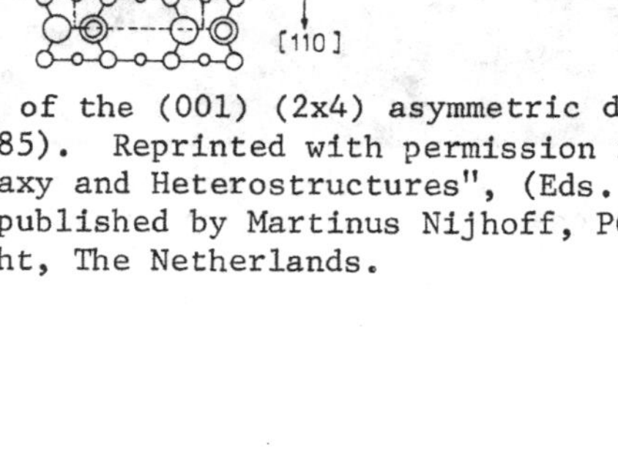

Fig. 5.16 Unit cell of the (001) (2x4) asymmetric dimer model (after
Joyce, 1985). Reprinted with permission from "Molecular
Beam Epitaxy and Heterostructures", (Eds. N Chang and
K Ploog) published by Martinus Nijhoff, PO Box 163, 3300
AD Dordecht, The Netherlands.

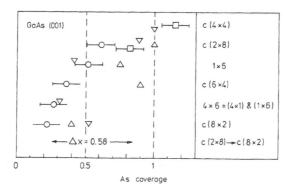

Fig. 5.17 Correlation between atomic arrangement and composition of
the GaAs (001) surface: △ van Bommel et al, 1978, Massies
et al, 1980; o Drathen et al, 1978; ▽ Bachrach et al, 1981);
△x Arthur, 1974 (after Mönsch, 1985). Reprinted with
permission from "Molecular Beam Epitaxy and Heterostructures", (Eds. N Chang and K Ploog) published
by Martinus Nijhoff, PO Box 163, 3300 AD Dordecht,
The Netherlands.

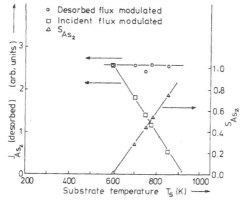

Fig.5.18 The sticking coefficient of As$_2$ on GaAs (001) as function
of temperature. The three lines show the relative
information available from modulating either the incident
or desorbing flux (after Joyce, 1985). Reprinted with
permission from "Molecular Beam Epitaxy and Heterostructures", (Eds. N Chang and K Ploog) published
by Martinus Nijhoff, PO Box 163, 3300 AD Dordecht,
The Netherlands.

Bachrach et al, 1981).   The change in structure with As coverage
(Monch, 1985) is shown in Figure 5.17 and indicates a transition
from the c(8x2) As-stabilised structure to a c(2x8)Ga stabilised
structure.   Massies et al (1980) noted that for As adsorption on an
As-stabilised surface (ie As coverage >1), a c(4x4) reconstruction
is observed.   Raising the temperature of GaAs produces the same
effect as moving from an As-stabilised surface to a Ga-stabilised
surface (Figure 5.18).   Later work by Larsen et al (1983) has shown
that the c(2x8) reconstruction is attributable to chemisorbed As
bonded to the As-stabilised surface.

Although we have discussed only the reconstruction of the (001)
GaAs surface, similar effects are observed on other surfaces and
have been widely studied (see Joyce, 1985; Monch, 1985). The
importance of such studies lies in establishing the foundation for
more detailed studies of the adsorption and surface-migration of
various precursor molecules, which will enable future crystal
growers to select the optimum precursor molecule for growth on a
specific surface.

CHAPTER 6

TRENDS IN CRYSTAL GROWTH

6.1 Introduction

In the preceding chapters we have seen how high growth temperatures lead to high point-defect and line-defect concentrations, impurity incorporation and diffusion. The convection effects associated with high-temperature growth lead to temporal and spatial non-uniformity in growth rate, whilst the broad-waveband radiation and the high temperature of the molecular species involved give rise to a multiplicity of chemical reactions and growth mechanisms. Theoretical analysis of the growth process is difficult if not impossible. For these reasons, early theories, which assumed for simplicity that growth occurred from the adsorption of a single precursor species, was kinetically limited by mobility on the growth surface and was unaffected by the desorption of product species, were of little practical significance. Growth of good quality single-crystals remained essentially an art rather than a science.

For some time now it has been recognised that the defects associated with high-temperature growth would be reduced by growth at lower temperatures, with a consequent increase in importance of liquid and vapour phase epitaxial techniques. As the growth temperature is lowered however, the surface mobility of the adsorbed species is reduced. This makes migration to satisfactory nucleation sites more difficult and promotes "island" growth with increased risk of defect generation due to discontinuity of the crystal lattice where adjacent islands meet. The reduced temperature also leads to an imperfect "lattice-filling sequence", whereby a large proportion of second-level sites are occupied before the first level is completely filled, resulting in vacancy and void formation. Desorption rates are also reduced, reducing the growth rate and

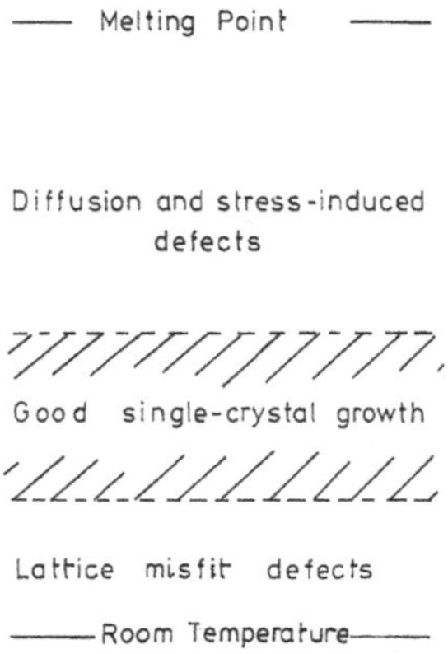

—— Melting Point ——

Diffusion and stress-induced
defects

Good single-crystal growth

Lattice misfit defects

——— Room Temperature———

Fig.6.1   The optimum temperature range for single-crystal growth.

leading to impurity incorporation through failure to remove the
reaction products. There is therefore a temperature "window"
(Figure 6.1), within which defect-free growth can be achieved. Note
that in Figure 6.1 the term "misfit defects" should not be confused
with "mismatch dislocations" which arise through the difference in
lattice parameter between the substrate and epilayer. The term
"misfit defect" refers to point, line or planar defects produced by
the incorrect location of atoms arising from any perturbation (eg
lattice damage, precipitation, mismatch dislocations etc). The low
growth temperature precludes its subsequent elimination by thermal
diffusion.

    Two significant recent developments have substantially altered
the existing situation in crystal growth technology and offer
exciting and potentially revolutionary prospects for the future. In
the first of these, techniques such as LPCVE and MBE have simplified
the deposition process by the removal of mass- transport effects and
the UHV conditions of MBE growth have facilitated studies of
adsorption-desorption reactions and growth kinetics. In the second,
the development of plasma, electron, ion and photon stimulation
techniques has made possible the injection of non-thermal energy
into the surface or near-surface regions of the growth system and
the modification of the energy state of the precursor species. The
immediate effect of the application of these techniques is to
increase our knowledge of the factors affecting the minimum growth
temperature. This is leading to significant reductions, which have
made possible the growth of very thin layers by reduction or
elimination of interface defects. These have been achieved in
practice using the rapid gas-switching facilities of MOVPE and MBE.
Further improvements in switching speed are emerging from the use of
plasma-controlled, electron-controlled or photon-controlled

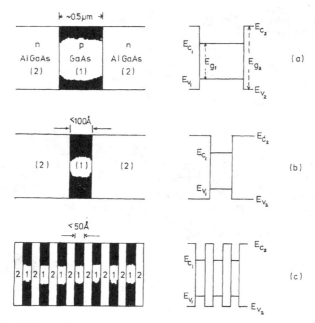

Fig.6.2   Schematic structure and energy band diagrams for
a) heterostructure, b) quantum well structure and
c) superlattice.

reactions. Such reactions are defined as those in which the
reaction rate falls to zero in the absence of the external
stimulation. This contrasts with plasma-enhanced reactions where
the effect of the plasma is simply to increase the reaction rate.
Additionally, photon- and electron-controlled reactions offer the
possibility of localised deposition in selected areas.

In the remainder of this chapter we shall examine first some of
the exciting new materials combinations which can be fabricated
using existing techniques and anticipate some of the developments in
crystal-growth techniques which will ultimately lead to very precise
growth control, selected area deposition and etching and the *in situ*
fabrication of complete multilayer, multi-material device structures.

6.2   Quantum Wells and Superlattices

The low-temperature kinetically-limited growth achievable by
MBE has made possible the production of atomically sharp
interfaces. In the lattice-matched $GaAs/Al_xGa_{1-x}As$ system
production of such interfaces resulted in the first efficient
heterojunction bipolar transistor, HJBT (Figure 6.2a) and

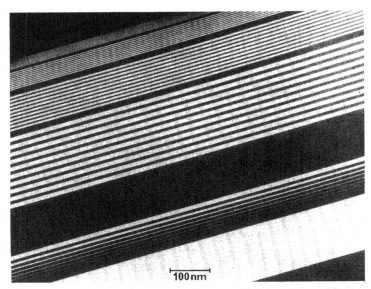

100 nm

Fig.6.3   Transmission electron micrograph of a cross-section
          of a GaAs/AlGaAs superlattice (courtesy of C Whitehouse,
          RSRE, Malvern).

double-heterostructure lasers. If the interface is sufficiently
sharp and the base region sufficiently narrow the base region
becomes a 'quantum well' (Figure 6.2b) in which electrons injected
from the emitter are confined and exhibit the properties of a
two-dimensional electron gas (2DEG). Such structures function as
high electron mobility transistors (HEMTs). In practical devices
several successive wide- and narrow-band-gap layers are deposited to
produce a multiple quantum well (MQW) structure. If the barrier
thickness between adjacent wells is reduced, electrons are no longer
confined, but can tunnel between them producing a materials
structure whose electronic and optical properties are different from
those of the end members and can be tuned by adjusting the applied
bias field. Such structures are referred to as 'superlattices'. A
typical example of a GaAs/AlGaAs superlattice is shown in Figure
6.3. The plethora of potential applications for quantum well and
superlattice structures has led to the development of a new field of
semiconductor physics under the general title of Low Dimensional
Solids (LDS). Recent reviews in this field include those by Gossard
(1981), Dohler (1983) and Dohler and Ploog (1984).

By far the most studied system to date is the GaAs/AlGaAs
combination where the lattice-mismatch is near zero, but
"strained-layer" superlattices have been produced in epitaxial
systems with large lattice mismatch, eg GaAs/InGaAs. The first
narrow band-gap II-VI superlattices were produced by Faurie et al

(1982). Recent reports on interdiffusion in this system (Arch et al, 1986; Zanio, 1986; Irvine et al, 1986) indicate that even at growth temperatures as low as 180°C interlayer diffusion is not entirely suppressed, suggesting that even lower growth temperatures are required for optimum electron confinement.

6.3 Heterostructures

The superlattice and quantum well structures discussed in section 6.2 are fabricated in the same alloy system. In parallel, however, rapid progress is now being made in the deposition of epitaxial layers whose composition is not related to that of the substrate (heteroepitaxy). The most studied system in this category is silicon on sapphire (SOS) (Cullen, 1978). Other examples include $Cd_xHg_{1-x}Te/CdTe/Al_2O_3$ (Tennant 1983), insulators on semiconductors, eg BaF_2 on InP (Sullivan et al, 1982) and CaF_2 on Si (Ishiwara et al, 1983) and metals on semiconductors, eg Al on GaAs (Cho and Dernier, 1978) and $CoSi_2$ on Si (Tung et al, 1982).

All these structures have been developed within the last decade and would not have been possible prior to the development of MBE and MOVPE low-temperature growth techniques. In these structures strains of up to 15% can be generated due to lattice mismatch. They can be accommodated elastically by tetragonal distortion in very thin layers or by localisation at the interface of the dislocation raft formed when such strains are relieved by plastic deformation. In high temperature growth processes the yield stress is too low to permit the former, whilst localisation of the dislocation array is prevented by cross-slip and climb. A second problem arising from high temperature growth lies in the fact that whilst lattice mismatch may be sufficiently small at the growth temperature,

Table 6.1 Lattice Parameters, Thermal Expansion and Elasticity Data for the $GaAs/SrF_2$ and $GaAs/CaF_2$ Systems (after Sullivan et al, 1985)

Material	Lattice Parameter $a(A)$	Thermal Expansion $\alpha(10^{-6}K^{-1})$	Elastic Constant c_{11}	$\frac{\Delta a}{a_0}$ 300K	$\frac{\Delta a}{a_0}$ 600K	$\frac{\Delta a}{a_0}$ 800K
				(w.r.t. a_0 GaAs)		
GaAs	5.6333	5.7				
SrF_2	5.7996	18.4	1247.5	2.56	3.02	3.41
CaF_2	5.464	19.2	1651.0	-3.40	-2.92	-2.48

raft formed when such strains are relieved by plastic deformation.
In high temperature growth processes the yield stress is too low to
permit the former, whilst localisation of the dislocation array is
prevented by cross-slip and climb. A second problem arising from
high temperature growth lies in the fact that although lattice
mismatch may be sufficiently small at the growth temperature,
thermal expansion differences between substrate and epilayer assume
increased importance when the growth temperature is significantly
above room temperature. This is well illustrated for the case of
CaF_2 and SrF_2 growth on GaAs (Sullivan et al, 1985). Reference to
Table 6.1, taken from Sullivan et al (1985), shows that the lattice
mismatch CaF_2/GaAs is similar to that of SrF_2 but the thermal
expansion coefficient is larger and the elastic modulus higher. In
consequence, CaF_2 films grown on GaAs tend to crack during cooling
from growth temperatures in the range 250-500°C (Farrow et al,
1985). SrF_2, despite not being prone to crazing like CaF_2, has
inferior electrical properties and is also inferior as a substrate
for GaAs growth. More recent research (eg Siskos et al, 1984) is
aimed at MBE growth of CaF_2-SrF_2 or CaF_2-BaF_2 alloys of better
lattice-match to GaAs.

The aim of such research is to grow both insulator/semi-
conductor and semiconductor/insulator structures, preferably at
similar temperatures, so that MBE or CVE techniques can be utilised
to produce stacked Metal-Insulator-Semiconductor (MIS) structures
for rapid parallel processing. Interconnections to the various
levels would be made by 'via hole' technology. Such structures are
sometimes referred to as 'high-rise' circuits by analogy with
multi-storey buildings. Even the achievement of the three level MIS
structure makes possible *in situ* 'chip' growth in a single growth
system. Heteroepitaxial growth of different semiconductors such as
GaAs on Si, to combine the laser and electro-optic switching
capabilities of the former with Si signal-processing is attracting

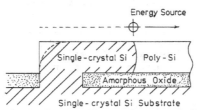

Fig.6.4   ELO - Epitaxial Layer Overgrowth of single-crystal
          silicon over an amorphous oxide film. Growth is seeded
          at the window in the oxide film and propagated
          laterally over the oxide surface using a moving hot-wire
          or focussed optical source.

Table 6.2   Summary of Heterostructure Growth

Substrate	Metal	Insulator	Semicon-ductor	Reference
Si	$CoSi_2$			Tang et al 1982
	"			Saitoh et al 1980*
	"			Kao et al 1985
	"			Arnaud et al 1985
		Al_2O_3/MgO		Ihara et al 1982*
		ZrO_2		Manasevit et al 1983
		CaF_2	Si	Ishiwara et al 1983*
				Fathauer et al 1985
		LaF_3		Sinharoy et al 1985
			GaAs	Uppal and Kromer 1985
			GaP	Uppal and Kromer 1985
			GaAs	Wang 1985
			GaAlAs	Wang 1985
			GaAs/CdTe	Bean et al 1986
GaAs	Al			Tung et al 1982(a)
	Fe			Kaplan and Bottka 1982
	"			Quadri et al 1985
		SrF_2	GaAs	Sullivan 1984
		$Sr_xCa_{1-x}F_2$	GaAs	Siskos et al 1984
		CaF_2		Sullivan et al 1985
			ZnSe	Park et al 1985
			Ge	Krautle et al 1983
			CdTe	Mullin et al 1981
				Feldman et al 1986
			HgTe-CdTe	Leopold et al 1986
			$Cd_xHg_{1-x}Te$	Nishitani et al 1983
				Giess et al 1985
InP	Al			Houzay et al 1985
		$Ba_xCa_{1-x}F_2$		Sullivan et al 1982
		"	InP	Tu et al 1983, 1983a
		CaF_2		Farrow et al 1981
				Sullivan et al 1982

(Continued)

* Multilayer structure

Table 6.2 cont'd.

Substrate	Metal	Insulator	Semiconductor	Reference
Al_2O_3			CdTe	Mullin et al 1981
			Si	Cullen 1978
			CdTe	Tennant 1983
				Cole et al 1984
			$Cd_xHg_{1-x}Te$	Tennant 1983
CaF_2			GaP	Cho 1970
InSb	Sn			Manasevit et al 1983
			CdTe	Farrow et al 1985
			CdTe	Williams et al 1985
			$Cd_xHg_{1-x}Te$	Noreika et al 1986
Ge		CaF_2	Si	Asano and Ishiwara 1982
			GaAs/AlGaAs	Wang 1985
$MgAl_2O_4$			CdTe	Mullin et al 1981

considerable research at the present time. The extent of multilayer heterostructure research is well illustrated by the recent survey made by Shaw (1983) and by the summary in Table 6.2.

6.4 Lateral Overgrowth

Perhaps even more challenging than the growth of widely lattice-mismatched epilayers on single crystal substrates is the attempted growth of good single-crystal layers of semiconductors over polycrystalline or amorphous substrates. Much of the research in this field is aimed at the growth of silicon over amorphous insulators such as SiO_2 and Si_3N_4, in an attempt to produce three-dimensional IC structures. In the technique first described by Jastrzebski et al (1982, 1983) and referred to as Epitaxial Lateral Overgrowth (ELO) a window is opened in the SiO_2 insulator to expose the underlying silicon single crystal and CVE deposition of Si from $SiCl_4$ is used to initiate growth at the window. Various techniques to promote lateral overgrowth have been adopted and some of these are reviewed in a special issue of J Crystal Growth on Single Crystal Silicon on Non-single-crystal Insulators (1983). A typical structure is illustrated in Figure 6.4.

From the above examples we see that the trend in modern crystal growth for semiconductor device fabrication is towards

a) low temperature growth to minimise diffusion

b) defect-free layer structures <1 μm in thickness

c) heterogeneous combinations of metals, insulators and semiconductors

In the remainder of the chapter we shall examine some of the developments in growth techniques which are leading to their achievement.

6.5 Chemical Beam Epitaxy

The growth of P_2-containing compounds by MBE from elemental P_2 sources is made difficult by the allotropic nature of the material

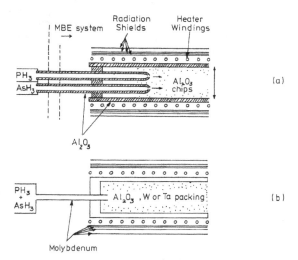

Fig.6.5 Schematic diagram of possible gas-source cells for chemical beam epitaxy. a) High pressure source, b) low pressure source (after Panish et al, 1985). Reprinted with permission from J.Vac.Sci.Technol. published by the American Institute of Physics, New York.

(resulting in irreproducible vapour pressures) and by the difficulties in achieving the necessary temperature control in the large-volume Knudsen cells required for P_2 (and As_2) generation. It is not surprising, therefore, that several groups of MBE workers have sought to replace elemental phosphorus and arsenic with gaseous sources (Panish, 1980; Cho and Chai, 1983; Kapitan et al, 1984). Two routes are available (Panish et al, 1985), as illustrated in Figure 6.5. In the first a high pressure reaction at 400-1500 Torr is used to produce M_4 where M is P or As:

$$4MH_3 \rightarrow M_4 + 6H_2 \qquad (6.1)$$

The M_4 is then passed through a high temperature cracking zone to produce M_2

$$M_4 \rightarrow 2M_2 \qquad (6.2)$$

Alternatively, a low-pressure gas-source is used, the PH_3 being passed over a metal catalyst, usually Ta (Panish et al, 1985; Calawa, 1981), to decompose the hydride. Growth rates are found to be faster (typically x2) than conventional MBE and the resulting GaAs has been used in the fabrication of state-of-the-art laser structures for 1.5 μm emitters.

Other workers (Veuhoff et al, 1981; Putze et al, 1985) have used the alternative approach of replacing the group III metal source by a high vapour-pressure metalorganic, eg $(CH_3)_3Ga$. (For this reason the technique is sometimes referred to as MOMBE (Metalorganic MBE). Unlike conventional MBE, where the growth rate

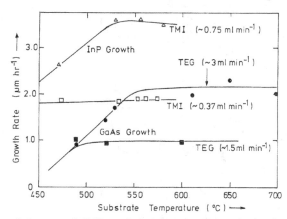

Fig.6.6   Growth rate of InP and GaAs as a function of substrate temperature and precursor species (after Tsang, 1985). Reprinted with permission from J.Vac.Sci.Technol. published by the American Institute of Physics, New York.

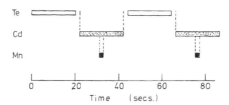

Fig.6.7   Switching sequence ('on' time) for component sources in
          the ALE growth of $Cd_{0.85}Mn_{0.15}Te$ (after Herman et al,
          1984). Reprinted with permission from J.Cryst.Growth,
          published by North Holland Publishing Co., Amsterdam.

is controlled by Ga adsorption, TMG is adsorbed only in the presence
of excess As. A potential problem of MOMBE is the incorporation of
high carbon levels due to incomplete pyrolysis and desorption of the
alkyl products. Putze reported carbon contents in the range
10^{19}-$10^{20}cm^{-3}$ for material grown from TMG, but noted a decrease of
several orders of magnitude when $(C_2H_5)_3Ga$, TEG, was used as the
precursor. Recently, Tsang (1985) introduced a completely organic
route to the preparation of InP, GaAs and $In_{0.53}Ga_{0.47}As$ using
$(CH_3)_3As$ and $(C_2H_5)_3P$ with $(CH_3)_3In$ and $(C_2H_5)_3Ga$. His results
again indicate the different reaction mechanisms occurring in MBE,
MOVPE and MOMBE. In MBE the growth- rate is determined by the Ga
beam flux; in MOVPE by the diffusion rate of the group III precursor
through the boundary layer and in MOMBE by the efficiency of
$(CH_3)_3Ga$ dissociation at the substrate surface.

At temperatures below 550°C the growth rate falls rapidly
(Figure 6.6) due to incomplete dissociation of $(C_2H_5)_3Ga$ at the flow
rate of 3 ml min^{-1}. The rate of decrease indicates an activation
energy of ~50 Kcal $mole^{-1}$, in good agreement with the heat of
formation of $(C_2H_5)_3Ga$ of 48 Kcal $mole^{-1}$. Reducing the flow rate to
1.5 cc min^{-1} leads to complete dissociation, even at temperatures
<500°C. For $(CH_3)_3In$ the temperature required for complete
dissociation at a flow rate of 0.75 ml min^{-1} is 550°C.

6.6   Atomic Layer Epitaxy

Although low-temperature growth techniques, particularly MBE,
afford some simplification of the growth process, the growth of
III-V and II-VI compounds still depends upon careful control of the
cation:anion flux arriving at the growth surface. A recent modification of MBE, atomic layer epitaxy (ALE) (Suntola, 1977; Suntola
et al, 1980; Goodman and Pessa, 1983), overcomes this problem
by using consecutive rather than simultaneous molecular beams of the
constituent elements. The duration of the first beam pulse is
sufficient to deposit one atomic layer of one of the constituents
(eg Cd in the case of CdTe growth) onto the growth surface. A

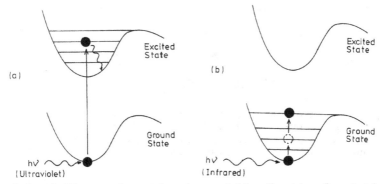

Fig.6.8   a) Electronic excitation and b) vibrational-rotational
          excitation due to photon absorption in the UV or
          IR wavelength ranges respectively.

second pulse from the Te source deposits a monolayer of Te.
Interdiffusion produces the required alloy. A dead time of 1-2
seconds between pulses ensures the removal of excess atoms not
adsorbed onto the growth surface. For non-stoichiometric or ternary
alloy growth, the required composition is achieved by adjusting the
duration of the individual precursor pulses. A typical sequence for
the growth of $Cd_{0.85}Mn_{0.15}Te$ (Herman et al, 1984) is shown in Figure
6.7. Potential disadvantages of the technique are the complexity
and cost of the switching operations required and the slow growth
rate ($\sim 1\,\text{\AA sec}^{-1}$) but the technique will undoubtedly find increasing
application over the next few years in fundamental studies of
sorption and surface migration rates.

6.7   Photochemical Deposition

     We have seen from preceding chapters that modern,
low-temperature growth of epitaxial structures requires the
localised absorption of energy by the precursor species in the
immediate vicinity of the substrate surface, providing only
sufficient energy to ensure the optimum adsorbed species, their
migration to suitable nucleation sites and the efficient removal of
waste products from the growth surface. In polycrystalline and
amorphous systems considerable success has been achieved by the use
of plasma-assisted and ion-beam techniques. The latter have also
been used in epitaxial growth (Itoh and Nakamura, 1971),
particularly for Si (Alexandrov et al, 1979), but reports on the
crystal quality suggest mixed success and it is unlikely that the
high energy (1-15keV) bombardment process would be applicable to
III-V or II-VI materials with their substantially lower damage
thresholds. For low-temperature, damage-free deposition,
photon-assisted processes are increasingly important.

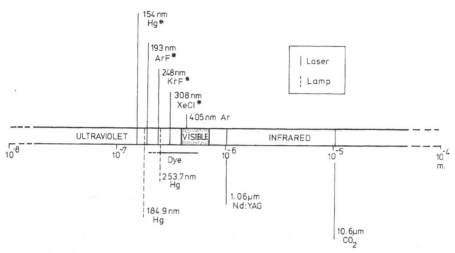

Fig.6.9   Laser and lamp source wavelengths commonly used in
           semiconductor photo-processing

We must differentiate at the outset between two essentially
different processes; photon-induced pyrolysis and photolysis. In
the former the radiation lies in the IR region of the spectrum and
is absorbed by the substrate causing the surface temperature to
rise, with consequent thermal cracking (pyrolysis) of the precursors. In photolysis, the energy is absorbed by the precursor
species. The absorbed photon may promote electrons to excited
states (Figure 6.8a) if UV or visible wavelengths are used, or
increase the ground state energy by single- or multi-photon
absorption processes at longer wavelengths (Figure 6.8b).

One of the problems associated with photolytic processes is
that of premature homogeneous reaction in the gas phase leading to
reaction products in the form of fine dust, which deposits on
reaction chamber walls and windows and on the substrate surface in
amorphous or polycrystalline state. Such deposits are virtually
impossible to recrystallise satisfactorily. Provided, however, that
satisfactory heterogeneous reaction at the substrate surface can be
achieved, photochemical processing offers several important advantages, including low-temperature selected-area deposition, reaction
control by optical modulation and, perhaps most importantly from the
standpoint of fundamental crystal growth studies, the possibility of
controlled dissociation to produce known precursor species whose
adsorption and mobility can then be studied under closely-controlled
experimental conditions. Figure 6.9 shows the most frequently used
radiation sources. Selected-area deposition may be achieved either

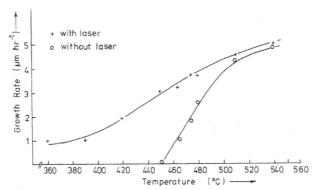

Fig.6.10 Growth rate of GaAs as a function of temperature for
conventional thermal deposition and for laser-assisted
deposition (λ = 530nm) (after Roth et al, 1983). Reprinted
by permission of the author and publisher from MRS Symp.
Proc. 17:193. Copyright 1985 by Elsevier Science Publishing Co. Inc., New York.

by projection imaging through out-of-contact masks (Ehrlich and
Tsao, 1984) employing UV or visible lamp sources for their high
intensity and good spatial uniformity, or by scanning laser sources
to produce 'line-writing' (Bauerle, 1983).

Although photochemical deposition has been extensively used to
produce metal line structures (Osgood and Ehrlich, 1982; Aylett and
Haigh, 1984; Adams et al, 1984) and large-area oxide and nitride
layers (Emery et al, 1984; Boyd 1984), its application to the
deposition of semiconductor materials is in its infancy. One of the
earliest reports of epitaxial growth was of (100) Ge film deposited
on (100) NaCl (Eden et al, 1983). Deposition was by photolytic
dissociation of GeH$_4$ in a He carrier gas to produce atomic Ge under
irradiation from a KrF (248mm) source. At the same time Roth et al
(1983) demonstrated laser-stimulated growth of GaAs at temperatures
as low as 360°C - well below the 450°C minimum observed in
conventional pyrolytic MOVPE deposition (Figure 6.10). In this case
the radiation was derived from a frequency-doubled Nd-YAG laser
(530nm) with a 3ns pulse width and 10Hz pulse repetition frequency
and it is highly probable that the dominant reaction was laser-
induced pyrolysis due to localised substrate heating.

Recent advances have included the deposition of InP (Donnelly
et al, 1984; Haig, 1985) and the II-VI compounds HgTe (Irvine et al,
1984) CdTe (Kisker et al, 1985) and Cd$_x$Hg$_{1-x}$Te (Irvine et al,
1985). The HgTe growth was the first reported instance of the UV
photolytic deposition of a compound semiconductor. An illustration
of the potential usefulness of photochemical techniques in

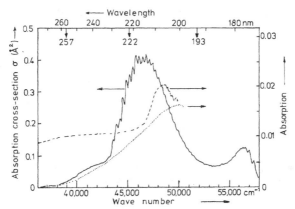

Fig.6.11 Absorption spectra for Cd $(CH_3)_2$; ___gas phase;
.... chemisorbed on SiO_2; ---- after irradiation with
a 193 mn laser beam (after Chen and Osgood, 1983).
Reprinted by permission of the authors and publisher
from MRS Symp.Proc. <u>17</u>:169. Copyright 1983 by Elsevier
Science Publishing Co. Inc., New York.

fundamental studies is given by the work of Chen and Osgood (1983)
and Irvine et al (1985a) on the dissociation of $Cd(CH_3)_2$. Figure
6.11 shows the UV absorption spectrum observed for vapour phase Cd
$(CH_3)_2$. The spectrum shows three main features: the transitions
indicated by the weak absorption feature on the long wavelength side
of the main peak result in the formation of a metal atom in the
ground state and two methyl radicals:

$$Cd(CH_3)_2 \overset{h\upsilon}{\to} Cd(^1s_o) + 2CH_3 \qquad (6.3)$$

The strong central peak is attributed to the formation of a methyl
radical and an excited metal methyl species which dissociates after
radiative emission to the ground state:

$$Cd(CH_3)_2 \overset{h\upsilon}{\to} Cd(^3p_1)CH_3 + CH_3 \qquad (6.4)$$

The short wavelength feature is associated with the formation of an
excited state metal atom and two CH_3 radicals:

$$Cd(CH_3)_2 \overset{h\upsilon}{\to} Cd(^3p_1) + 2CH_3 \qquad (6.5)$$

By judicious tuning of the incident photon energy it may be

Table 6.3   Surface Diagnostic Techniques Employed in Crystal
            Growth

	Technique	Basis	Reference
RHEED	Reflexion High Energy Electron Diffraction	Grazing angle Diffraction of ~5 KeV electrons from substrate or layer surface	Cho, 1971
AES	Auger Emission Spectroscopy	Electron bombardment causes characteristic Auger emission from adsorbed impurity and native atoms	Ploog and Fischer 1977
MS	Mass Spectrometry	Velocities of ionised impurity or native species from the growth surface are measured in a 'time of flight' spectrometer	
MMS	Modulated Mass Spectrometry	The incident molecular beam of an MBE process is modulated and the detector phase-locked to differentiate between beam and background or adsorbing and desorbing species	Neave and Joyce 1978
LIF	Laser-Induced Fluorescence	A tunable UV radiation source is used to promote spontaneous emission from gas phase species	Karlicek et al 1983
LIPE	Laser-Induced Photo-fragment Emission	Similar to LIF but in this case the absorbed radiation leads to photon-induced molecular fragmentation and characteristic emission	Karlicek et al 1983

(Continued)

Table 6.3 continued

	Technique	Basis	Reference
LRS	Laser Raman Scattering	The gas phase is irradiated with a fixed frequency source and the Raman shift characteristic of gas phase moecules is monitored spectroscopically	Kappitz et al 1984
PAS	Photo-Acoustic Spectroscopy	Irradiation with 2800-3800cm^{-1}UV generates an acoustic wave through thermal expansion associated with gas phase absorption. This is detected by a surface mounted piezoelectric transducer	Brueck et al 1983
SE	Ellipsometry	The change in polarisation of a reflected monochromatic light beam is a function of surface layer refractive index and is therefore a sensitive guide to changes in surface composition and structure	Theeten and Hottier, 1979
SHG	Second Harmonic Generation	532 nm laser radiation is reflected from the growth surface and the second harmonic radiation generated by interaction between the incident beam and the surface is monitored using a photomultiplier. The SHG output is sensitive to surface electronic properties determined by the adsorbed species	Tom et al, 1984

possible to derive different reaction species and study their adsorption and migration characteristics.

The dotted curve in Figure 6.11 represents the absorption spectrum obtained when the $Cd(CH_3)_2$ molecule is chemisorbed onto a silica surface. The absorption peak becomes much broader and is shifted towards 50.000 cm^{-1}, which Chen and Osgood interpret as a result of the reaction of $Cd(CH_3)_2$ with the OH radical in the SiO_2 surface to produce $Cd(^1p_1)CH_3$. Irradiation with 193 nm UV from an ArF excimer laser produces the third spectrum shown in Figure 6.11, with enhanced absorption in the 30,000-40,000 cm^{-1} range. This difference in absorption spectrum between gas-phase and surface-adsorbed species may well prove to be the key to successful suppression of unwanted gas-phase reactions and the promotion of heterogeneous, surface-specific growth reactions. For this reason photochemical deposition is attracting increasing attention as a research field and is prompting the development of surface diagnostic techniques.

6.8 Surface Diagnostic Techniques

Over the last decade MBE UHV surface analysis techniques such as RHEED, AES and SIMS have contributed enormously to our knowledge of the fundamental processes involved in crystal growth. Unfortunately such techniques are not applicable to CVE processes operating in the 1-760 Torr pressure range. Recently there has been increasing interest in the development of laser probe techniques which, when successfully accomplished, will confer on CVE processes the same degree of *in situ* monitoring and growth control as is currently available in MBE. A comprehensive review is beyond the scope of this book but a short summary of some of these emerging techniques is given in Table 6.3. Many of the techniques are not new and several have been reviewed by Omenetto (1979). What is new is their application to and optimisation for semiconductor growth studies.

6.9 Conclusion

Time and space do not permit a more detailed review of developments in the art and science of crystal growth. I hope, however, that the information contained in these pages is sufficient to demonstrate the tremendous changes now occurring in both the theory and practice of crystal growth. Techniques such as MBE and MOVPE and their derivatives are milestones on the pathway towards the crystal grower's ultimate aim - the deposition of the optimum precursor species on a correctly prepared substrate, with just sufficient energy to ensure migration to the correct nucleation site for incorporation in the perfect crystalline lattice. Even the imperfect achievement of some of these goals is leading inexorably towards thin, multilayer, selected area, heterogeneous structures, which promise to revolutionise optical and electronic signal processing techniques in the next decade.

REFERENCES

* Selected References Abstracted

Abe, T., Kikuchi, K., Shirai, S., Murdoka, S., 1981
    Semiconductor Silicon, p.55. Eds. H.Huff, Kreigler and
    Takeishi. Electrochem.Soc., Pennington, N.J.
* Adams, A.E., Lloyd, M., Morgan, S.L., Davis, N.G., 1984
    'Laser photochemical deposition of metals' in 'Laser
    Processing and Diagnostics', D. Bauerle, Ed., Springer Series in
    Chem.Phys. 39:269, Berlin 1984
Akiyama, M., Kawarada, Y., Kaminishi, K., 1984, Jap.J.
    Appl.Phys. 23(11):L843
* Aleksandrov, L.N., Lutovich, A.S., Belorusets, E.D.,
    1979, Phys.Stat.Sol. (a) 54:463
* Arch, D.K., Faurie, J.P., Chow, P., Staudermann, J-L,
    Hibbs-Brenner, M., 1986, J.Vac.Sci.Technol., A4(4):2101
Asano, T., Ishiwara, H., 1982, Jap.J.Appl.Phys.21:1630,
* Arnaud d'Avitaya, F., D.Delage, S., Rosencher, E., Derien,J.,
    1985, J.Vac.Sci.Technol. B3(2):770
* Astles, M., Gordon, N.T., Bradley, D., Dean. P.J., Wight,
    D.R., 1984, J.Electron.Mater. 13:167
Astles, M., Blackmore, G., Steward, V., Rodway, D.C.,
    Kirton, P., 1987, J.Cryst. Growth 80:1
* Augustus, P.D., Stirland, D.J., 1980, J.Microscopy 118:111
* Augustus, P.D., Stirland, D.J., 1982, J.Electrochem.Soc.
    129:614
* Aylett, M.R., Haigh, J., 1984, 'Laser photolytic deposition
    of metals on InP' in 'Laser Processing and Diagnostics',
    D Bauerle, Ed., Springer Series in Chem. Phys. 39:263,
    Berlin 1984
* Bachrach, R.Z., Bauer, R.S., Chiaradia, P., Hansson,
    G.V., 1981, J.Vac.Sci.Technol. 19:335
* Baluffi, R.W., 1980, High angle grain-boundaries as sources
    or sinks for point defects p297 in "Grain Boundary
    Structures and Kinetics", H.Gleiter, Ed., ASM Metals Park, Ohio
    44073
* Baluffi, R.W., Brokman, A., King, A.H., 1982, Acta.Met.
    30:1453
* Ban, V.S., 1978, J.Electrochem.Soc. 125:317
Bardsley, W., Cockayne, B., Green, G.W., Hurle, D.T.J.,
    Joyce, G.C., Roslington, J.M., Tufton, P.J., Webber, H.C.,

1974, J.Cryst.Growth 24/25:369

* Bardsley, W., Green, G.W., Holliday, C.H., Hurle, D.T.J.,
  Joyce, G.C., MacEwan, W.R., Tufton, P.J., 1975, in "GaAs
  and Related Compounds, 1974". (Proc. Vth Int.Symp., Deauville)
  Inst.Phys.Conf.Ser. 24:355

* Barraclough, K.G., 1982, Control of foreign point defects
  during Czochralski silicon crystal growth, in "Aggregation
  Phenomena of Point Defects in Silicon", E.Sirtl and J.
  Goorisen, Eds., Electrochem.Soc., Pennington N.J.

* Barraclough, K.G., Series, R.W., 1985, Oxygen Content of
  n^+ and p^+ Czochralski silicon, in "Proc. Symp. on Low-
  temperature Processing", Electrochem.Soc. Fall Meeting

* Barnett, S.J., Tanner, B.K., Brown, G.T., 1985, in "Advanced Photon
  and Particle Techniques for the Characterisation of Defects in
  Solids", Roberto, J.B., Carpenter, R.W., Wittels, M.C., Eds.
  Mat.Res.Soc.Symp.Proc. 41:83

  Bass, S.J., 1975, J.Cryst.Growth, 31:172

* Basson, J.H., Booyens, H., 1983, Phys.Stat.Sol. (a) 80:663

* Bauerle, D., 1983, Mat.Res.Soc.Symp.Proc. 17:19

  Bell, R.L., Willoughby, A.F.W., 1966, J.Mater.Sci. 1:219

* Bean, R., Zanio, K., Hay, K., Wright, J., 1986, J.Vac.Sci.
  Technol., A4(4):2153

  Bell, R.L., Willoughby, A.F.W., 1966. J.Mater.Sci. 1:219

* Bell, S.L., Sen, S., 1985, J.Vac.Sci.Technol. A3(1):112

* Beneking, H., Escobosa, A., Krautle, H., 1981, J.Electron.Mater.
  10(3):473

  Bennema, P., Gilmer, G.H., 1973, Kinetics of crystal growth,
  in "Crystal Growth: an Introduction", P.Hartman, Ed., North
  Holland, Amsterdam

* Bensahel, D., Magnea, N., Pautrat, J.L., Pfister, J.C.,
  Revoil, L., 1979, Defects and Radiation Effects in Semi-
  conductors, Proc.Int.Conf., Nice, Sept. 1978. Inst.Phys.Conf.
  Ser. 46:421

  Berkman, S., Ban, V.S., Goldsmith, N., 1977, in "Heteroepitaxial
  Semiconductors for Electronic Devices", G.W.Cullen and C.C.
  Wang, Eds., Springer Verlag, New York

  Bethe, H.A., 1935, Proc.Roy.Soc. A150:552

  Blunt, R.T., Clark, S., Stirland, D.J., 1982, IEEE Trans.
  Microwave Theory Tech. 30:943

  de Boer, J.H., 1968, "The Dynamical Character of Adsorption",
  Clarendon Press, Oxford

  Bollman, W., 1970, "Crystal Defects and Crystalline Interfaces"
  Springer Verlag, New York

  Booker, R., Lyster, M., Oxford University - unpublished work

* Bourret, A., Thibault-Dessaux, J., Seidman, D.N., 1984,
  J.Appl.Phys. 55:825

* Boyd, I.W., 1984, in "Laser Processing and Diagnostics",
  D.Bauerle, ed., Springer Verlag, Berlin

* Bozler, C.A., 1975, J.Electrochem.Soc. 122(12):1705

  Brice, J.C., 1965, "Growth of Crystals from the Melt", Vol. 5,

Selected Topics in Solid State Phys. P.Wohlfarth Ed. North Holland, Amsterdam

Brice, J.C., 1973, "Growth of crystals from liquids, Vol.12, 'Selected Topics in Solid State Physics, P.Wohlfarth, Ed., North Holland, Amsterdam

Brice, J.C., 1986, 'Crystal growth processes', Blackie Halstead Press, John Wiley, New York

Van der Brekel, C.H.J., 1978, Ph.D.Thesis, University of Nijmegen, The Netherlands

Bridgman, P.W., 1925, Proc.Amer.Acad.Arts and Sci., 60:305

* Brown, G.T., Cockayne, B., MacEwan, W.R., 1983, J.Electron. Mater. 12:93

Bunn, C., 1949, Disc.Faraday Soc., 5:144

Burton, W.K., Cabrera, N., Frank, F.C., 1951, Phil.Trans. Roy.Soc. A243:299

Burton, J.A., Prim, R.C., Schlichter, W.P., 1953, J.Chem.Phys., 21:1987

* Cadoret, R., Cadoret, M., 1975, J.Cryst.Growth, 31:142

* Calawa, A.R., 1981, App.Phys.Lett., 38:701

* Capper, P., 1982, J.Cryst.Growth, 57:280

* Capper, P., Gosney, J.J., Jones, C.L., 1984, J.Cryst.Growth, 70:356

* Capper, P., Gosney, J.J., Jones, C.L., Kenworthy, I., Roberts, J.A. 1985, J.Cryst.Growth, 71:57

Carlson, A.E., 1958, in "Growth and Perfection of Crystals", R,.H.Doremus, B.W.Roberts, D.Turnbull, eds., Wiley, New York, Chapman and Hall, London

Carruthers, J.R., 1967, J.Electrochem.Soc., 114:959

Carruthers, J.R., Nassau, K., 1968, J.Appl.Phys. 39:5205

* Carruthers, J.R., 1976, J.Cryst. Growth. 32:13

Chakraverty, B.K., 1973, 'Heterogeneous nucleation and condensation in crystal growth: an introduction', P. Hartman, ed., North Holland, Amsterdam

Chalmers, B., Gleiter, H., 1971, Phil.Mag. 23:1541

Chamonal, J.P., 1982, Ph.D. Thesis, University of Grenoble

* Chang, C.E., Wilcox, W.R., 1974, J.Cryst. Growth, 21:135

* Chang, H-R., 1985, J.Electrochem.Soc. 132(1):219

Chang, L.L., and Ploog, K., 1985, 'Molecular beam epitaxy and heterostructures',, NATO ASI Series E: App.Sci. No.87, Martinus Nijhoff, Dordecht, Boston, Lancaster

Chedzey, H., Hurle, D.T.J., 1966, Nature, 239:993

* Chen, C.J., Osgood, R.M., 1983, Mater.Res.Soc.Symp.Proc. 17:169 in 'Laser diagnostics and photochemical processing for semiconductor devices', R.M.Osgood, S.R.J.Brueck, H.R. Schlossberg, Eds.

* Chen, K., 1984, J.Cryst. Growth, 70(1/2):64

* Chen, R.T., Holmes, D.E., 1983, J.Crystal Growth, 61:111

Chew, N.G., Williams, G.M., Cullis, A.G., 1984 in 'Electron Microscopy and Analysis', P.Doig, Ed., Inst.Phys., Bristol

* Chew, N.G., Cullis, A.G., Warwick, C.A., Robbins, D.J.,

Hardeman, R.W., Gasson, D.B., 1985, 'Electron microscope characterisation of silicon layers grown by molecular beam epitaxy' in 'Microscopy of Semiconducting Materials', A.G.Cullis and D.B. Holt, Eds., Inst.Phys.Conf.Series, 76:129

* Cho, A.Y., 1970, J.Appl.Phys. 41:782

* Cho, A.Y., 1971, J.Vac.Sci.Technol., 8:531

Cho, A.Y., Arthur, J.R., 1975, 'Progress in Solid State Chemistry' Somorjai, G. and McCaldin, J., Eds. 3:73, Springer-Verlag, Berlin.

* Cho, A.Y., Dernier, P.D., 1978, J.Appl.Phys. 49:3328

* Cho, A.Y., 1985, 'Growth and properties of III-V semiconductors by molecular beam epitaxy' p191 in 'Molecular Beam Epitaxy and Heterostructures', L.L. Chang and K. Ploog, Eds., NATI ASI Series E; Appl,.Sci. No.87, Martinus Nijhoff, Dordecht, Boston, Lancaster

* Chow, R., Chai, Y.G., 1983, J.Vac.Sci.Technol. 1:49

* Claeys, C., Bender, H., Declerck, G., van Landuyt, J., van Overstraeten, R., Amelinckx, S., 1982, 'Kinetics of process -induced defect complexes in silicon' in 'Aggregation Phenomena of Point Defects in Silicon', E.Sirtl and J. Goorisen, eds., Electrochem.Soc., Pennington, N.J.

Cockayne, B., Gates, M.P., 1967, J.Mater.Sci. 2:118

* Cockayne, B., Roslington, J.M., 1973, J.Mater.Sci. 8:601

* Cockayne, B., Brown, G.T., MacEwan, W.R., 1981, J.Cryst. Growth, 51:461

* Cockayne, B., Brown, G.T., MacEwan, W.R., 1983, J.Cryst. Growth, 64:48

Cole, H.S., Woodbury, H.H., and Schetzina, J.F., 1984, J. Appl.Phys. 55:3166

* Cole, S., 1982, J.Cryst. Growth 59:370

* Coltrin, M.E., Kee, R.J., Miller, J.A., 1984, J.Electrochem. Soc., 131:425

Cooper, C.B., Ludowise, M.J., Albi, V., Moon, R.L., 1980, J.Electron.Mater. 9:299

* Crochet, M.J., Wouters, P.J., Geyling, F.T., Jordan, A.S., 1983, J.Cryst. Growth, 65:153

* Crochet, M.J., Geyling, F.T., Van Schaftingen, J.J., 1983, J.Cryst. Growth, 65:166

Cullen, G.W., 1978, 'Heteroepitaxial Semiconductors for Electronic Devices', G.W.Cullen and C.C.Wang, Eds., Springer, New York

* Cullen, G.W., Duffy, M.T., Jastrzebski, L., Lagowski, J., 1983, J.Cryst. Growth, 65:415

* Cullis, A.G., Katz, L.E., 1974, Phil.Mag. 30:1419

* Cullis, A.G., Augustus, P.D., Stirland, D.J., 1980, J.Appl. Phys., 51(5):2556

Curtis, J., 1917, Z.Physik.Chem. (Leipzig), 92:219

* Curtis, B.J., Dismukes, J., 1972, J.Cryst.Growth, 17:128

Dash, W.C., 1956, J.Appl.Phys. 27:1193

Davies, G.J., Williams, D., 1985, 'III-V MBE Systems' in 'The Technology and Physics of Molecular Beam Epitaxy', E.H.C Parker, Ed,, Plenum Press, New York

Dismukes, J.P., Curtis, B,J., 1973, in 'Semiconductor Silicon', H.R.Huff and R,R, Burgess, Eds., Electrochem.Soc., New York

* Dobson, P.S., Fewster, P.F,, Hurle, D.T.J., Hutchinson, P.W., Mullin, J.B., Straughan, B.W., Willoughby, A.F.W., 1979, in "GaAs and Related Compounds" (Proc. VIIth Int.Symp., St. Louis, Missouri, Sept. 1978), Inst.Phys.Conf.Ser. No.45:16.

* Dobson, P.J., Neave, J.H., Joyce, B.A., 1982, Surf.Sci. 119:L339

Dohler, G.H., 1983, Scientific American 11:118

Dohler, G.H., Ploog, K., 1984, Adv. in Phys.

* Donahue, T.J., Burger, W.R., Reif, R., 1984, Appl.Phys.Lett., 44(3):346

* Donnelly, V.M., Geva, M., Long, J., Karlicek, R.F., 1984, Appl.Phys.Lett. 44:951

* Drathen, P., Ranke, W., Jacobi, B., 1978, Surf.Sci. 77:L162

* Duchemin, J-P., Bonnet, M., Koelsch, F., 1978, J.Electrochem. Soc. 125:637

* Duchemin, J-P., 1981, J.Vac.Sci.Technol. 18(3):753

Dushman, S., 1962, in 'Scientific Foundations of Vacuum Technique', 2nd Ed. J.M.Lafferty, ed., Wiley, New York

* Eden, J.G., Greene, J.E., Osmundsen, J.F., Lubben, D., Abele, C.C., Gorbatkin, S., Desai, H.D., 1983, in 'Laser Diagnostics and Photochemical Processes for Semiconductor Devices', R.M.Osgood, S.R.J.Brueck, H.R. Schlossberg, Eds., Mat. Res.Soc.Symp.Proc. 17:185

* Ehrlich, D.J., Tsao, J.Y., 1984, in 'Laser-controlled Chemical Processing of Surfaces', A.W.Johnson, D.J.Ehrlich, H.R.Schlossberg, Eds., Mat.Res.Soc.Symp.Proc., 29:195

Elwell, D., Scheel, H.J., 1975, 'Crystal Growth from High-temperature Solutions', Academic Press, London

Emery, K., Boyer, P.K., Thompson, L.R., Solanki, R., Zarnani, H., Collins, G.J., 1984, Proc. SPIE, 459:9

* van Erk, W., van Hoek-Martens, H.J.G.J., Bartels, G., 1980, J.Cryst.Growth, 48:621

* Eversteyn, F.C., Severin, P.J., van der Brekel, C.H.J., Peek, H.L., 1970, J.Electrochem.Soc. 117:925

Farrar, R.A., Gilham, C.J., Bartlett, B., Quelch, M.J., 1977, J.Mat.Sci. 12:836

Farrow, R.F.C., 1974, J.Phys.D: Appl.Phys. 7:2436

* Farrow, R.F.C., Sullivan, P.W., Williams, G.M., Jones, G.R.,. Cameron, D.C., 1981, J.Vac.Sci.Technol. B19:415

* Farrow, R.F.C., Wood, S., Greggi, J.C.Jr., Takei, W.J., Shirland, F.A., Furneaux, J., 1985, J.Vac.Sci.Technol. B3(2):681

* Fathauer, R.W., Lewis, N., Showalter, L.J., Hall, E.L., 1985, J.Vac.Sci.Technol. B3(2):736

* Faurie, J-P., Million, A., Piaguet, J., 1982, Appl.Phys.Lett. 41:713

* Feldman, R.D., Kisker, D.W., Austin, R.F., Jeffers, K.G.,
    Bridenbaugh, P.M., 1986, J.Vac.Sci.Technol., A4(4):2234
* Fink, H.W., Ehrlich, G., 1981, Surf.Sci. 110:L611
  Fink, H.W., 1982, Ph.D.Thesis, Munich
* Fischer, R., Henderson, T., Klem, J., Masselink, W.T., Kopp, W.,
    Morkoc, H., 1984, Electron.Lett. 25(22):945
  Fowler, 1962, in 'Introduction to Statistical Mechanics',
    T.L.Hill, Ed., Addison, Wiley, New York
  Foxon, C.T., Harvey, J.A., Joyce, B.A., 1973, J.Phys.Chem.Sol.
    34(3):1693
  Foxon, C.T., Joyce, B.A., 1977, Surf.Sci. 64:293
  Frank, F.C., 1975, Comments on nucleation theory p1 in
    'Crystal Growth and Characterisation', R.Ueda and J.B.
    Mullin, Eds., North Holland, Amsterdam
* Fujii, E., Nakamura, H., Haruna, K., Koga, Y., 1972, J.Electrochem.
    Soc., 119:1106
* Giess, J., Gough, J.S., Irvine, S.J.C., Blackmore, G.W.,
    Mullin, J.B., Royle, A., 1985, J.Cryst.Growth 72:120
* Giling, L.J., 1982, J.Electrochem.Soc., 129(3):634
* Gilmer, G.H., Broughton, J.Q., 1983, J.Vac.Sci.Technol.,
    B1(2):298
  Gossard, A.C., 1982, 'Molecular beam epitaxy of superlattices
    in thin films' in 'Treatise on Materials Science and
    Technology', 24:13, Academic Press
  Griffiths, R.J.M., Chew, N.G., Cullis, A.G., Joyce, G.C.,
    1983,, Electron.Lett., 19:988
  Gross, U., Kersten, R., 1972, J.Cryst.Growth, 15:85
* Haigh, J., 1985, J.Vac.Sci.Technol., B3(5):1456
  Halicioglu, T., 1980, Phys.Stat.Sol.(b) 99:347
* Hardeman, R.W., Robbins, D.J., Gasson, D.B., Daw, A., 1985, Proc.1st
    Int.Symp. on Silicon MBE. 167th ECS Meeting, Toronto, Canada
* Hartman, P., Perdock, W.G., 1955, Acta Cryst. 8:49,521
  Hartman, P., 1973, "Structure and Morphology" in 'Crystal Growth:
    an Introduction', P. Hartman, Ed., North Holland, Amsterdam, p.3
    67
* Herman, M.A., Jylha, O., Pessa, M., 1984, J.Cryst.Growth, 66:480
* Hersee, S.D., Duchemin, J.P., 1982, Ann.Rev. Mater.Sci. 12:65
* Hersee, S.D., Baldy, M., Duchemin, J.P., 1983, J.Electron.Mater.
    12(2):345
  Hirsch, P.B., Howie, A., Nicholson, R.B., Pashley, D.W., Whelan,
    M.J., 1965, 'Electron Microscopy of Thin Crystals', Butterworths
    London
* Hitchman, M.L., 1980, J.Cryst.Growth 48:394
  Hitchman, M.L., Curtis, B.J., 1981, 'Prog. in Cryst.Growth and
    Characterisation'. 4:285
* Hitchman, M.L., Curtis, B.J., 1982, J.Cryst.Growth 60:43
  Hoerni, J.A., 1960, IRE Electron. Devices Meeting, Washington
* Holmes, D.E., Chen, R.T., Elliott, K.R., Kirkpatrick, C.G.,
    1983, Appl.Phys.Lett. 43:305
  Holt, D.B., 1966, J.Phys.Chem.Sol. 27:1053

Hoshikawa, K., Hirata, H., Nakanishi, H., Ikuta, K., 1981, in
    'Semiconductor Silicon' (Proc. 4th Int.Symp.on Silicon Materials
    Science and Technology). H.Huff, R.J.Kriegler and Y.Takeishi
    Eds. p.101, Electrochem.Soc., Pennington, N.J.
* Hottier, F., Theeten, J.B., 1980, J.Cryst.Growth 48:644
Hougen, O.A., Watson, K.M., 1947, in 'Chemical Process Principles
    III - Kinetics and Catalysis', John Wiley and Sons, New York
* Houzay, F., Moison, J.M., Bensoussan, M., 1985, J.Vac.Sci.
    Technol., B3(2):756
Hu, H., 1972, 'The Nature and Behaviour of Grain Boundaries',
    Plenum, New York
Hudson, J.B., Sandejas, J.S., 1967, J.Vac.Sci.Technol. 4/5:230
Hudson, J.B., Sandejas, J.S., 1968, Surf.Sci. 5:125
Huff, H.R., Kriegler, R.J., Takeishi, Y., 1981, (Eds) Semiconductor
    Silicon, Electrochem.Soc., Pennington, N.J.
Hull, D., 1965, in 'Introduction to Dislocations' p.158, Pergamon,
    Oxford
Hurle, D.T.J., 1967, in 'Crystal Growth', H.S.Peiser, Ed.,
    Pergamon, Oxford
Hurle, D.T.J., Jakeman, E., Johnson, C.P., 1974, J.Fluid.Mech.
    64:565
Hurle, D.T.J., 1976, 'Hydrodynamics of Crystal Growth' in "Current
    Topics in Materials Science', Vol.2, E.Kaldis and H.J.Scheel,
    Eds., North Holland Publishing Co., Amsterdam, New York, Oxford
* Hurle, D.T.J., 1978, J.Phys.Chem.Sol. 40:613. ibid 627(Te);
    ibid 639 (Sn); ibid 647 (Ge)
* Hurle, D.T.J., 1980, J.Cryst.Growth 50:638
* Hurle, D.T.J., 1983, J.Cryst.Growth 65:124
* Hwang, G.T., Cheng, K.C., 1973, J.Heat Transfer Trans. ASME 95:72
* Ihara, M., Animoto, Y., Jifuku, M., Kimura, T., Kodama, S.,
    Yamawaki, H., Yamaoka, T., 1982, J.Electrochem.Soc. 129:2569
* Irvine, S.J.C., Mullin, J.B., Tunnicliffe, J., 1984, J.Cryst.
    Growth 68:188
* Irvine, S.J.C., Giess, J., Mullin, J.B., Blackmore, G.W.,
    Dosser, O.D., 1985, J.Vac.Sci.Technol. B3:1450
* Irvine, S.J.C., Mullin, J.B., Robbins, D.J., Glasper, J.L., 1985a,
    J.Electrochem.Soc. 132(4):968
* Irvine, S.J.C., Giess, J., Gough, J.S., Blackmore, G.W., Royle, A.,
    Mullin, J.B., Chew, N.G., Cullis, A.G., 1986, J.Cryst.Growth
    77:437
* Ishiwara, H., Asano, T., Furukawa, S., 1983, J.Vac.Sci.Technol.
    B1:266
Itoh, T., Nakamura, T., 1971,, Rad.Effect 9:1
* Jackson, K.A., 1958, 'Mechanism of Growth' in 'Liquid Metals and
    Solidification', ASM, Cleveland, 174
* Jackson, K.A., 1975, Theory of melt growth, p21 in 'Crystal Growth
    and Characterisation', R. Ueda and J.B. Mullin, Eds., North
    Holland
Jacob, G., Duseaux, M., Farges, J.P., van den Boom, M.M.B.,
    Roksneer, P.J., 1983, J.Cryst.Growth 61:417

Jakeman, E., Hurle, D.T.J., 1972, Rev.Phys.Technol. 3:3

* James, T.W., Stoller, R.E., 1984, Appl.Phys.Lett. 44(1):56

* Jastrzebski, L., Corboy, J.F., Pagliaro, R., 1982, J.Electrochem. Soc. 129:2645

* Jastrzebski, L., 1983, J.Cryst.Growth 63:493

Jones, C.L., Quelch, M.J.T., Capper, P., Gosney, J.J., 1982, J. Appl.Phys. 53(12):9080

* Jones, C.L., Capper, P., Gosney, J.J., 1982, J.Cryst.Growth 56(3): 581

* Jones, C.L., Capper, P., Gosney, J.J., Kenworthy, I., 1984, J. Cryst.Growth 69:281

* Jordan, A.S., 1980, J.Cryst.Growth, 49(4):631

Joyce, B.A., 1974, Rep.Prog.Phys. 37:363

Joyce, B.A., Foxon, C.T., 1975, J.Cryst.Growth 31:122

* Joyce, B.A., Foxon, C.T., Neave, J.H., 1978, J.Jap.Assoc.Cryst. Growth 5:185

* Joyce, B.A., 1985, 'Kinetic and surface aspects of MBE'p37 in 'Molecular Beam Epitaxy and Heterostructures', L.L. Chang and K.Ploog, Eds., NATO ASI Series E - Applied Sciences No. 87, Martinus Nijhoff, Dordecht, Boston, Lancaster

* Juza, J., Cermak, J., 1982, J.Electrochem.Soc. 129:1627

Kakutani, T., 1962, J.Phys.Soc.Jap. 17:1496

Kaldis, E., 1974, 'Principles of vapour growth of single crystals' in 'Crystal Growth, Theory and Techniques', Vol. 1, C.H.L. Goodman,Ed., p.49, Plenum Press, London, New York

* Kao, Y.C., Tejwani, M., Xie, Y.H., Lin, T.L., Wang, K.L., 1985, J.Vac.Sci.Technol. B3(2):596

* Kapitan, L.W., Litton, C.W., Clarke, G.C., Cottier, P.C., 1984, J.Vac.Sci.Technol. B2:280

* Kaplan, R., Bottka, N., 1982, Appl.Phys.Lett. 41:972

* Karlicek, R.F., Donnelly, V.M., Johnston, W.D., 1983, 'Laser spectroscopic investigation of gas-phase processes relevant to semiconductor device fabrication' MRS Symp.Proc. 17:151

* Kawaguchi, Y., Asaki, H., Nagai, H., 1985, Jap.J.Appl.Phys. 24(4):L221

* Kisker, D.W., Feldman, R.D., 1985, J.Cryst.Growth 72:102

* Kobayashi, N., 1981, J.Cryst.Growth 52:425

de Koch, A.J.R., 1973, 'Microdefects in Dislocation-free Silicon Crystals', Ph.D. Thesis, Catholic University of Nijmegen

de Koch, A.J.R., 1976, 'Effect of growth conditions on silicon semiconductor quality' in 'Current Topics in Materials Science II', E.Kaldis, Ed., North Holland, Amsterdam

* Kolbesen, B.O., 1982, 'Carbon in silicon' in 'Aggregation Phenomena of Point Defects in Silicon', E.Sirtl and J.Goorisen, Eds., p.155, Electrochem.Soc., Pennington, N.J.

* Koppitz, M., Vestavik, O., Mircea, A., Heyen, M., Richter, W., 1984, J.Cryst.Growth 64:136

* Krautle, H., Roentgen, P., Beneking, H., 1983, J.Cryst.Growth 65(1-3):439

Kroger, F.A., Vink, H.J., 1956, in 'Solid State Physics', F.Seitz and D.Turnbull, Eds., p.310. Academic Press, New York

Kronberg, M.L., Wilson, F.H., 1949, Trans. AIME, 185:151

Kuhlmann-Wilsdorf, D., Wilsdorf, H.G.F., 1960, J.Appl.Phys. 31:516

Kunzel, H., Knecht, J., Jung, H., Wunstel, K., Ploog, K., 1982, Appl.Phys. A28:167

Kyropoulus, S., 1926, Z.Anorg.Chem. 154:308

Laister, D., Jenkins, G.M., 1983, J.Mat.Sci. 8:1218

* Lagowski, J., Gatos, H.C., Parsey, J.M., Wada, K., Kaminska, M., Walukiewicz, W., 1982, Appl.Phys.Lett. 40:342

* Lagowski, J., Lin, D.G., Aoyama, T., Gatos, H.C., 1984, Appl.Phys. Lett. 44:336

Landau, A.I., 1958, Phys.Met.Metallog. 6:148

Langlois, W.E., 1981, Physico-Chem.Hydrodynam. 2:245

* Langlois, W.E., 1984, J.Cryst.Growth, 70:73

* Larsen, P.K., Neave, J.H., van der Veen, J.F., Dobson, P.J., Joyce, B.A., 1983, Phys.Rev. B27:4966

* Lee, H.H., 1984, J.Cryst.Growth, 68:698

Leopold, D.J., Wroge, M.L., Ballingall, J.M., Morris, B.J., Peterman, D.J., Broerman, J.G., 1986, J.Vac.Sci.Technol. B4(6):1306

* Lodge, E.A., Booker, G.R., Warwick, C.A., Brown, G.T., 1985, Inst. 'TEM study of grown-in dislocations in semi-insulating undoped LEC GaAs' in 'Microscopy of Semi-conducting Materials', A.G.Cullis and D.B.Holt, Eds., Inst.Phys.Conf.Series 76:217

Lorimer, G.W., 1975, J. de Physique 36:233 (Supp.Coll.C4)

Lunn, M., 1986, J.Cryst.Growth - to be published

MacRae, A.U., Gobeli, G.W., 1966, in 'Semiconductors and Semi-metals' Vol. 2, R.K.Willardson and A.C.Beer, Eds., p.115, Academic Press, New York

Magee, T., Peng, J., Bean, J., 1975, Phys.Stat.Sol.(a) 27:557

* Maguire, J., Newman, R.C., Beall, R.B., 1986, J.Phys.C. 19:1897

Manasevit, H.M., 1968, Appl.Phys.Lett., 12:158

Manasevit, H.M., Simpson, W.I. 1973, J.Electrochem.Soc. 120:137

* Manasevit, H.M., Golecki, I., Moudy, L.A., Yang, J.J., Mea, J.E., 1983, J.Electrochem.Soc. 130:1752

* Mandel, G., 1964, Phys.Rev. 134(4A):A1073

* Manke, C.W., Donaghey, L.F., 1977, J.Electrochem.Soc. 124:561

Marfaing, Y., 1981, Prog.Cryst.Growth and Charact. 4:317

* Martin, G.M., 1981, Appl.Phys.Lett., 39:747

* Masselink, W.T., Fischer, R., Klem, J., Henderson, T., Morkov, H., 1985, J.Vac.Sci.Technol. B3(2):548

* Massies, J., Etienne, P., Dezaly, F., Linh, N.T., 1980, Surf.Sci. 99:121

Matsumura, T., Emori, H., Terashima, K., Fukuda, T., 1983, Jpn. J.Appl.Phys. 22:L154

* Matsuo, S., Kiuchi, M., 1983, Jpn.J.Appl.Phys., 22(4):L210

Matthews, J.W., Mader, S., Light, t.B., 1970, J.Appl.Phys. 41:3800

Matthews, J.W., 1975, 'Epitaxial Growth', Academic Press, New York

Matthews, J.W., 1979, 'Misfit dislocations' in 'Dislocations in Solids', F.R.N.Nabarro, Ed., North Holland, Amsterdam

Mattis, D.C., 1965, in 'The Theory of Magnetism', Harper and Row, New York

McFee, J.M., Miller, B.I., Bachmann, K.J., 1977, J.Electrochem.Soc. 124(2):253

* Miheličic, M., Wingerath, K., 1985, J.Cryst.Growth, 71:163

* Mirsky, U., Scechtman, D., 1980, J.Electron.Mater. 9(6):933

Mimura, T., Hiyamizu, S., Fujii, T., Nambu, K., 1980, Jpn.J.Appl. Phys. 19:L225

* Mönch, W., 1985, in 'Molecular Beam Epitaxy and Heterostructures', A.Y.Cho and K.Ploog, Eds. NATO ASI Series E - Applied Sciences No. 87, Martinus Nijhoff, Dordecht, Boston, Lancaster

Morgan, A.E., 1974, J.Cryst.Growth 27:226

* Moss, R.H., 1983, J.Cryst.Growth, 65:463

* Müller-Krumbhaar, H., 1974, Phys.Rev.B No.4 10:1308

Müller-Krumbhaar, H., 1976, 'Developments in the theory of the roughening transition' in 1976 'Crystal Growth and Materials', E.Kaldis and H.J.Scheel, Eds., North Holland, Amsterdam

Mullin, J.B., Hulme, K.F., 1960, J.Phys.Chem.Sol. 17:1

Mullin, J.B., Royle, A., Straughan, B.W., 1970, Proc.3rd Int.Conf. on GaAs and Related Compounds: Aachen. Inst.Phys.Conf.Series, 9(1972):L331

* Mullin, J.B., Irvine, S.J.C., Ashen, D.J., 1981, J.Cryst.Growth, 55:92

* Mullin, J.B., Irvine, S.J.C., 1981, J.Phys.D: Appl.Phys. 14:2149

* Mullin, J.B., Irvine, S.J.C., Royle, A., Tunnicliffe, J., Blackmore, G.W., Holland, R., 1983, J.Vac.Sci.Technol. A1:1612

Mullens, W.W., Sekerka, R.F., 1964, J.Appl.Phys. 35:444

Mutaftschiev, B., 1965, 'Adsorption et Croissance Cristalline', Centre National de Recherche Scientifique, Paris, p.231

Mutaftschiev, B., 1983, J.Cryst.Growth, 65:50

* Nagao, S., Higashitani, K., Akasaka, Y., Nakata, H., 1985, J. Appl.Phys. 57(10:4589

Nanishi, Y., Ishida, S., Miyazawa, S., 1983, Jpn.J.Appl.Phys. 22:L54

* Neave, J.H., Joyce, B.A., 1978, J.Cryst.Growth, 44:387

Neave, J.H., Blood, P., Joyce, B.A., 1980, Appl.Phys.Lett. 36:311

Nernst, W., 1904, Z.Phys.Chem. 47:52

Newman, R.C., Woodhead, J., 1984, J.Phys.C: Sol.State Phys: 17:1405

Newman, R.C., 1985, in 'Festkorperprobleme', Adv.in Sol.State Phys. XXV Vieweg, Braunschweig

Nguyen, A., Cinti, R., Chakraverty, B.K., 1972, J.Cryst.Growth 13/14:174

* Nishitani, K., Ohkata, R., Murotani, T., 1983, J.Electron.Mater. 12(3):619

* Noreika, A.J., Farrow, R.F.C., Shirland, F.A., 1986, J.Vac.Sci. Technol. A4(4):2081

* Normand, C., Pomeau, Y., Velarde, M.G., 1977, Rev.Mod.Phys. 49(3):581

* Olsen, G.H., 1975, J.Cryst.Growth, 31:225
* Olson, J.M., Rosenberger, F., 1979, J.Fluid Mech. 92(4):609
  Omenetto, N., 1979, 'Analytical Laser Spectroscopy', Wiley and
    Sons, New York
  Onsager, L., 1944, Phys.Rev. 65:117
* Osgood, R.M., Ehrlich, D.J., 1982, Opt.Lett., 7:385
* Pande, K.P., Seabaugh, A.C., 1984, J.Electrochem.Soc. 131:1357
* Panish, M., 1980, J.Electrochem.Soc., 127:2729
* Panish, M., Temkin, H., Sumski, S., 1985, J.Vac.Sci.Technol.
    B3(2):657
* Park, R.M., Mar, H.A., Salansky, N.M., 1985, J.Vac.Sci.Technol.
    B3(2):676
  Parker, E.H.C.,, 1985, 'The Technology and Physics of Molecular
    Beam Epitaxy', Plenum, New York, London
  Patel, J.R., Chaudhuri, A.R., 1963, J.Appl.Phys. 34:2788
* Pautrat, J.L., Magnea, N., Faurie, J.P., 1982, J.Appl.Phys.
    53(12):8668
  Pearson, E., Takei, T., Halicioglu, T., Tiller, W.A., 1984, J.
    Cryst.Growth, 70:33
* Penning, P., 1958, Philips Res.Repts., 13:79
* Pessa, M., Huttunen, P., Herman, M.A., 1983, J.Appl.Phys.
    54(10):6047
  Petroff, P., Kimmerling, L.C., 1976, Appl.Phys.Lett., 29:461
* Phillips, J.M., Gibson, J.M., 1984, 'The growth and character-
    isation of epitaxial fluoride films on semiconductors'p381 in
    'Thin Films and Interfaces II', J.E.E.Baglin, D.R. Campbell and
    W.K. Chu, Eds., MRS Symp.Proc. 25:381
  Plaskett, T.S., 1965, Trans. AIME 223:809
  Ploog, K., Fischer, A., 1977, Appl.Phys. 13:111
* Ploog, K., 1980, 'Molecular beam epitaxy' in "Crystals, Growth,
    Properties and Applications', Vol.3:73, Springer Verlag, Berlin,
    Heidelberg, New York
* Ponce, F.A., Yamashita, T., Hahn, S., 1983, Appl.Phys.Lett.,
    43:1051
  Ponce, F.A., Hahn, S., Yamashita, T., Scott, M., Carruthers, J.R.,
    1983, in 'Microscopy of Semiconducting Materials', A.G.Cullis,
    S.M.Davidson and G.R.Booker, Eds., Inst.Phys.Conf.Series, 65
* Ponce, F.A., 1985, 'Structure of micro-defects in semiconducting
    materials' in 'Microscopy of Semiconducting Materials', A.G.
    Cullis, D.B.Holt, Eds., Inst.Phys.Conf.Series, 76:1
* Pond, R.C., Gowers, J.P., Holt, D.B., Joyce, B.A., Neave, J.H.,
    Larsen, P.K., 1984, Mat.Res.Soc.Symp.Proc., 25:273
* van der Putte, Giling, L.J., Bloem, J., 1975, J.Crys.Growth,
    31:299
* Putz, N., Veuhoff, E., Heineeke, H., Heyen, M., Luth, H.,
    Balk, P., 1985, J.Vac.Sci.Technol. B3(2):671
* Qadri, S.B., Goldenberg, M., Prinz, G.A., Ferrari, J.M., 1985,
    J.Vac.Sci.Technol. B3(2):718
* Queisser, H.J., 1983, 'Electrical properties of dislocations and
    boundaries in semiconductors' in 'Defects in Semiconductors',

Vol.2:323, S.Mahagan and J.W.Corbett, Eds., North Holland.

* Razeghi, M., Poisson, M.A., Larivain, J.P., Duchemin, J.P., 1983,
  J.Electron.Mater. 12(2):371.

* Reif, R., Dutton, R.W., 1981, J.Electrochem.Soc. 128(4):909

* Robbins, D.J., Pidduck, A.J., Hardeman, R.W., Gasson, D.B.,
  Pickering, C., Daw, A.C., Chew, N.G., Cullis, A.G., Johnson, M.,
  Jones, R., 1987, J.Cryst.Growth, 81:421

* Roberts, J.S., Dawson, P., Scott, G.B., 1981, Appl.Phys.Lett.
  38(11):905

* Robertson, D.S., 1966, Brit.J.Appl.Phys. 17:1047

* Roth, W., Krautle, H., Krings, A., Beneking, H., 1983, MRS Symp.
  Proc. 17:193

* Rozgonyi, G.A., Deysher, R.P., Pearce, C.W., 1976, J.Electrochem.
  Soc. 123:1910

* Russell, G.J., Waite, P., Woods, J., Lewis, K.L., 1981, in
  2nd Microscopy of Semiconducting Materials Meeting, Oxford

  Saitoh, S., Ishiwara, H., Furukawa, S., 1980, Appl.Phys.Lett.
  37:203

  Salermo, J.P., Gale, R.P., Fan, J.C.C., 1981, 'Scanning cathodo-
  luminescence microscopy of polycrystalline GaAs' in 'Defects in
  Semiconductors' p59, Narayan and Tan, Eds., North Holland,
  Amsterdam

  Salin, F.A., Swaminathan, V., Kroger, F.A., 1975, Phys.Stat.Sol.
  (a)29:465

* Schaake, H.F., Tregilgas, J.H., Lewis, A.J., Everett, P.M., 1983,
  J.Vac.Sci.Technol. A1(3):1625

* Schaake, H.F., Tregilgas, J.H., 1983, J.Electron.Mater.
  12(6):931

* Schaake, H.F., Tregilgas, J.H., Beck, J.D., Kinch, M.A.,
  Gnade, B.E., 1985, J.Vac.Sci.Technol. A3(1):143

* Scheel, H.J., Schultz-Dubois, E.O., 1971, J.Cryst.Growth 8:304

  Schlier, R.E., Farnworth, M.E., 1959. Chem.Phys. 30:917

  Schlichter, W.P., Burton, J.A., 1958, in 'Transistor Technology':
  107, H.E.Bridgers, J.H.Scaff, J.N. Schive,. Eds., van Nostrand,
  Princeton

  Schlichting, H., 1968, 'Boundary-layer theory',, 6th Ed.:177,
  McGraw Hill, New York

* Schmidt, P.F., Pearse, C.W., 1981, J.Electrochem.Soc. 128(3):630

* Seager, C.H., 1982, 'The electronic properties of semiconductor
  grain boundaries' p85 in 'Grain Boundaries in Semiconductors',
  Pike, C.H.Seager and Lemy, Eds., Elsevier Publishing Co.

* Series, R.W., Barraclough, K.G., 1982, J.Cryst.Growth, 60:212

* Series, R.W., Barraclough, K.G., 1983, J.Cryst.Growth, 63:219

  Series, R.W., Barraclough, K.G., Hurle, D.T.J., Kemp, D.S.,
  Rae, G.J., 1985, Extended Abstracts, Electrochem.Soc., Spring
  Meeting

  Shaw, D.W., 1968, J.Electrochem.Soc. 115:405

  Shaw, D.W., 1974, in 'Crystal Growth Theory and Techniques 1:1,
  C.H.Goodman, Ed., Plenum Press, New York, London

* Shaw, D.W., 1983, J.Cryst.Growth, 65(1/3):444

* Shimada, T., Terashima, K., Nakajima, H., Fukuda, T., 1984, Jpn.J.Appl.Phys. 23(1):L23

  Shiraki, Y., 1985, 'Silicon molecular beam deposition' in 'The Technology and Physics of Molecular Beam Epitaxy' p345, E.H.C. Parker, Ed., Plenum Press, New York, London

* Sinharoy, S., Hoffman, R.A., Rieger, J.H., Takei, W.J., Farrow, R.F.C., 1985, J.Vac.Sci.Technol. B3(2):722

  Sirtl, E., Adler, A., 1961, Z.Metallkunde, 52:529

  Sirtl, E., Goorissen, J., 1982, Eds., 'Aggregation Phenomena of Point Defects in Silicon', Electrochem.Soc., Pennington, NJ

* Siskos, S., Fontaine, C., Munoz-Yague, A., 1984, Appl.Phys.Lett. 44:146

* Skolnick, M.S., Brozel, m.R., Reed, L.J., Grant, I., Stirland, D.J., Ware, R.M., 1984, J.Electron.Mater. 13:107

  Smallman, R.E., 1962, 'Modern Physical Metallurgy', Butterworths, London

  Sparrow, E.M., Elchorn, R., Greg, J.L., 1959, Phys.Fluids, 2:319

  Stanley, C.F., Farrow, R.F.C., Sullivan, P.W., 1985, 'MBE of InP and other phosphorus-containing compounds' in 'Technology and Physics of MBE':275, E.H.C. Parker, Ed., Plenum Press, New York

  Stewart, M.J., Weinberg, F., 1972, J.Cryst.Growth, 12:217

* Stirland, D.J., Augustus, P.D., Straughan, B.W., 1978, J.Mater. Sci. 13:657

* Stirland, D.J., 1986, 'Direct observations of defects in semi-insulating GaAs' in 'Proc.4th Int.Conf. on Semi-insulating III-V Materials, Hakone, Japan, May 1986'

  Stockbarger, D.C., 1936, Rev.Sci.Instr. 7:133

* Sugawara, K., 1972, J.Electrochem.Soc. 119:1749

* Sullivan, P.W., Farrow, R.F.C., Jones, G.R., 1982, J.Cryst.Growth 60:403

* Sullivan, P.W., 1984, Appl.Phys.Lett. 44:140

* Sullivan, P.W., Bower, J.E., Metze, G., 1985, J.Vac.Sci.Technol. B(3):500

  Suntola, T., Antson, M.J., 1977, US Patent No. 4058 430

  Suntola, T., Antson, M.J., Pakkala, A., Lindfors, S., 1980, Soc. for Information Display Digest, p.108

  Takabatake, T., Furuya, T., Ueda, Y., 1970, Jpn.J.Appl.Phys. 9:416

  Taniguchi, M., Ikoma, T., 1982, in "GaAs and Related Compounds, 1982" (Proc.Int.Symp. Albuquerque, 1982). Inst.Phys.Conf.Series, 65:65

* Taniguchi, M., Ikoma, T., 1983, J.Appl.Phys., 54:6448

  Tempkin, D.E., 1966, in 'Crystallisation Processes', Consultants Bureau, New York

  Tempkin, D.E., 1971, Sov.Phys.Crystallog. 15:767

  Tennant, W.E., 1983, TEDM Tech.Dig:704

  Theeten, J.B., 1976, 'Growth mechanisms in CVD of GaAs' in '1976 Crystal Growth and Materials', E.Kaldis and H.J.Scheel, Eds., North Holland, Amsterdam

* Tiller, W.A., 1984, J.Cryst.Growth, 70:13
* Tom, H.W.K., Mate, C.M., Zhu, X.D., Crowell, J.E., Heinz, T.F.,
    Shen, Y.R., Phys.Rev.Lett., 52(5):348
  Townsend, W.G., Uddin, M.E., 1973, Sol.State Elect. 16:39
* Tregilgas, J., Beck, J., Gnade, B., 1985, J.Vac.Sci.Technol.
    A3(1):150
* Tsang, W.T., Miller, R.C., Capasso, F., Bonner, W.A., 1982,
    Appl.Phys.Lett., 41(5):467
* Tsang, W.T., 1985, J.Vac.Sci.Technol. B3(2):666
* Tu, C.W., Forrest, S.R., Johnston, W.D. Jr., 1983, Appl.Phys.
    Lett. 43:569
* Tu, C.W., Sheng, T.T., Read, M.H., Schlier, A.R., Johnson, J.G.,
    Johnston, W.D., Bonner, W.A., 1983a, J.Electrochem.Soc.
    130:2081
* Tung, R.T., Bean, J.C., Gibson, J.M., Poate, J.M., Jacobson, D.C.,
    1982, Appl.Phys.Lett. 40:684
* Tung, R.T., Poate, J.M., Bean, J.C., Gibson, J.M., Jacobson, D.C.,
    1982a, Thin Solid Films, 93:77
* Tunnicliffe, J., Blackmore, G.W., Irvine, S.J.C., Mullin, J.B.,
    Holland, R., 1984, Mat.Letts. 2(5a):393
  Tweet, A.G., 1959, J.Appl.Phys. 30:2002
  Uppal, P.N., Kromer, H., 1985, J.Vac.Sci.Technol. B3(2):603
  van Vechten, J.A., 1975, J.Electrochem.Soc. 122:423
  van Vechten, J.A., 1984, J.Phys.C., 17:L933
* Vere, A.W., Cole, S., Williams, D.J., 1983, J.Electrochem.Mater.
    12:551
* Vere, A.W., Steward, V., Jones, C.A., Williams, D.J., Shaw, N.,
    1985, J.Cryst.Growth, 72:97
* Veuhoff, E., Pletschen, W., Balk, P., Luth, H., 1981, J.Cryst.
    Growth, 55:30
  Volmer, M., 1939, 'Die Kinetik der Phasenbildung', Steinkopff,
    Dresden
* Wahl, G., 1977, Thin Solid Films, 40:13
* Wahl, G., Hoffman, R.H., 1980, Rev.Int.Haute Temp.Refract. 17:7
* Wang, W.I., 1984, Appl.Phys.Lett., 44(12):1149
  Wang, W.I., 1985, J.Vac.Sci.Technol. B3(2):552
* Wanklyn, B.M., 1983, J.Cryst.Growth, 65:533
  Wannier, G.H., 1945, Rev.Mod.Phys., 17:50
* Warrington, D.H., 1980, 'Formal geometric aspects of grain
    boundary structure':1 in 'Grain Boundary Structure and
    Kinetics', ASM Metals Park, USA
* Warwick, C.A., Brown, G.T., Booker, G.R., Cockayne, B., 1983,
    J.Cryst.Growth, 64:108
* Warwick, C.A., Brown, G.T., 1985, Appl.Phys.Lett., 46(6):574
* Weber, E.R., Ennen, H., Kaufmann, U., Windscheif, J., Schneider
    J., Wosinski, T., 1982, J.Appl.Phys., 53:6140
* Westphal, G.H., 1983, J.Cryst.Growth, 65:105
  Wheeler, A.A., 1984, J.Cryst.Growth, 67(1):8
  Wilkes, J.G., 1983, J.Cryst.Growth, 65:214
* Wilkes, J.G., 1984, J.Cryst.Growth, 70:271

Williams, D.J., 1986, Ph.D.Thesis, University of Birmingham, UK
* Williams, G.M., Whitehouse, C.R., Chew, N.G., Blackmore, G.W.,
    Cullis, A.G., 1985, J.Vac.Sci.Technol. B3(2):704
* Williams, D.J. and Vere, A.W., 1986, J.Vac.Sci.Technol., A4(4):2184
Williams, D.J., 1986, J.Cryst.Growth - in press
Witt, A.F., Gatos, H.C., 1968, J.Electrochem.Soc. 115:70
Wood, E.A., 1964, J.Appl.Phys., 35:1306
* Wright, P.J., Cockayne, B., 1982, J.Cryst.Growth, 59:148
* Wright, P.J., Griffiths, R.J.M., Cockayne, B., 1984, J.Cryst.
    Growth, 66:26
* Yoshida, M., Watanabe, H., Uesugi, F., 1985, J.Electrochem.Soc.
    132:677
* Zanio, K., 1986, J.Vac.Sci.Technol., A4(4):2106
Zinnes, A.E., Nevis, B.E., Brandle, C.D., 1973, J.Cryst.Growth
    19:187
Zou Yuanxi, 1982, in "GaAs and Related Compounds, 1981" (Proc.
    Int.Symp., Oiso, Japan). Inst.Phys.Conf.Series, 63:185
* Zulehner, W., 1983, J.Cryst.Growth, 65:189

SELECTED ABSTRACTS

Adams, A.E., Lloyd, M., Morgan, S.L. and Davis, N.G.

LASER PHOTOCHEMICAL DEPOSITION OF METALS

The paper describes the deposition of patterned W and WSi_2 films on quartz and silicon substrates by Laser Chemical Vapour deposition (LCVD) and Laser Photo deposition (LPD).

Tungsten films were deposited from WF_6 onto silicon at temperatures in the range 200-500°C. Sheet resistance was found to be dependent on deposition temperature, showing a minimum (3Ω/suare) at 400°C. Auger analysis showed traces of oxygen in the metal film. SEM micrographs indicate a columnar grain structure in the 70-100 nm thick films.

Tungsten silicide was deposited from a mixture of WF_6 and 10% silane in helium. At T >150°C a pyrolysed film containing both W and Si is obtained. At lower temperatures photolytic decomposition gives a W-deposit with little or no silicon. The nature of the possible photolytic reactions responsible for such observations is discussed.

'Laser Processing and Diagnostics', Ed. D.Bauerle, Springer Series in Chem.Phys. 39:269. Springer Verlag, Berlin, 1984.

Aleksandrov, L.N., Lutovich, A.S., and Belorusets, E.D.

THE MECHANISM OF SILICON EPITAXIAL LAYER GROWTH FROM ION-MOLECULAR BEAMS

Energies of different stages of silicon homoepitaxial growth from ion-molecular beam (IMB) in vacuum: oxide layer destruction, physical absorption and chemisorption, surface diffusion etc. are estimated. Based on the discussion of the values obtained conclusions on the possible mechanism of the ion beam influence on nucleation and epitaxial layer growth are made. In the places of ion collision with crystallizing surface, point defects and local areas of atom excitation are formed which become centres of nucleation. Ions colliding with the centres of growth destroy the volume nuclei providing the conditions of two-dimensional covering. Ion beam makes the energetic additional feeding of diffusion and phase transition processes.

Reprinted with permission from Phys.Stat.Sol.(a) 1979, 54:463

Arch, D.K., Faurie, J.P., Staudenmann, J-L., Hibbs-Brenner, M. and Chow, P.

INTERDIFFUSION IN HgTe-CdTe SUPERLATTICES

Semiconductor superlattices comprised of alternating HgTe and CdTe layers have been proposed as a tunable narrow band gap semiconductor for long wavelength optoelectronic applications. Growth of this novel superlattice material has been reported by several laboratories. Its usefulness as a narrow band gap optical material, however, has not been established. A main issue of concern is the interdiffusion of the constituent Hg, Cd and Te atoms across the heterointerfaces of the superlattice structure. To determine the extent of this interdiffusion we have carried out, for the first time, temperature dependent X-ray diffraction measurements on HgTe-CdTe superlattices. Peak intensities of the superlattice satellites were monitored as a function of annealing temperature and time to yield estimates of the interdiffusion coefficient $D(T)$.. Our results indicate there is appreciable intermixing of the HgTe and CdTe layers at temperatures as low as 110°C. Such results have serious implications for the use of this material in optoelectronic devices.

Reprinted with permission from J.Vac.Sci.Technol. 1986 (A4)(4):2101, published by The American Institute of Physics, New York. © 1986 American Vacuum Society.

Arnaud d'Avitaya, F.A., Delage, S., Rosencher, E., and Derrien, J.

KINETICS OF FORMATION AND PROPERTIES OF EPITAXIAL CoSi$_2$ FILMS ON Si (111)

Very thin CoSi$_2$ films, epitaxially grown on silicon (111) surfaces, have been obtained under ultrahigh vacuum conditions by thermal reaction of Co layers deposited onto Si (111) substrates. The morphology and structural properties of the $CoSi_2$ films depend strongly on experimental parameters such as film thickness, annealing temperature to form $CoSi_2$ or substrate temperature during Co deposition. Usually for high-temperature formation ($\geqslant700°C$) and large thickness ($\geqslant200A$) $CoSi_2$ islands are observed. Below these limits the $CoSi_2$ films display a rather smooth and homogeneous aspect. The reaction kinetics are also followed by *in situ* monitoring of the Co and Si Auger peaks versus annealing temperature. Two interfaces have been examined, namely Co-Si and Co-$CoSi_2$. Plateaus in the Auger peak variation show the evolution of the sampled region from the Co metallic phase to successively Co_2Si, $CoSi$ and $CoSi_2$ phases. Finally, electrical characteristics of some nearly perfect $CoSi_2$-Si Schottky barriers have been checked with intensity and capacitance versus voltage techniques. The deduced barrier ranges around $0.65\pm0.02eV$ and no signal has been detected with deep-level transient spectroscopy.

Reprinted with permission from J.Vac.Sci.Technol., 1985, 33(2):770, published by the American Institute of Physics. © 1985 American Vacuum Society.

Astles, M., Gordon, N., Bradley, D., Dean, P.J., Wight, D.R. and Blackmore, G.

DOPING OF LPE LAYERS OF CdTe GROWN FROM Te SOLUTIONS

Good quality epitaxial layers of CdTe have been grown by LPE from Te-rich solutions at ~500°C onto (111) CdTe substrates. The layers have been characterised by a wide variety of techniques including capacitance-voltage profiling, photoluminescence at 4K and secondary-ion mass spectrometry (SIMS). Undoped layers had good electrical properties ($p \sim 1 \times 10^{16}cm^{-3}$, $L_e \sim 3\mu m$) and SIMS showed the layers to be of high purity. Those doped with In and Aℓ, however, had low n-type carrier concentrations and very short diffusion lengths, while the photoluminescence spectra showed a strong peak at ~ 1.4eV commonly seen in n-type bulk CdTe. The most heavily doped layers showed marked decreases in lattice parameter.

These results are fully discussed and a model to explain them is outlined. The basis of this is that at the end of the LPE growth cycle, Cd is rapidly lost from the layer surface producing V_{Cd} centres which form deep-level complexes of the type $In_{Cd} \cdot V_{Cd}$.

From SIMS analysis of the grown layers, the distribution coefficients of the dopants In and Aℓ under the growth conditions used have been estimated as $k_{In} = 0.06$ and $k_{A\ell} = 0.28$.

Reprinted with permission from J.Electron.Mater. 1984, 13:167, published by the Metallurgical Society, Warrendale, PA, USA.

Augustus, P.D. and Stirland, D.J.

A STUDY OF INCLUSIONS IN INDIUM PHOSPHIDE

A detailed examination of unusual inclusions in InP substrates has been undertaken. Techniques employed have included optical microscopy, scanning electron microscopy, reflection and transmission X-ray topography, conventional and high voltage electron microscopy, electron diffraction, energy dispersive X-ray analysis, and Auger emission spectroscopy. Inclusions have been detected in many, but not all, undoped, Fe-, Sn-, and Ge-doped InP ingots, at densities from ~10^3 to 10^4 cm^{-2}. Larger inclusions consist essentially of a central core of densely tangled dislocations, surrounded by dislocation loops predominantly on {110} planes and extending 10-20 μm outwards from the core. The entire complex of core plus surrounding loops results in a characteristically pitted appearance after etching, or as a region of strong contrast by X-ray topography. Smaller inclusions (~1 μm size) have been examined in electron microscope specimens foils. Some inclusions have been found at line dislocations, and it is probable that this is where the majority of inclusions originate, by a nucleation process. Attempts to determine the structure and chemical composition of the inclusions are presented and discussed.

Reprinted with permission from J.Electrochem.Soc. 1982, 129:614, published by The Electrochemical Society.

Augustus, P.D., and Stirland, D.J.

MICROSCOPY OF SEMI-INSULATING GALLIUM ARSENIDE

Optical and electron microscopy examinations of a wide range of semi-insulating, n-type and (one) p-type single crystal (001) gallium arsenide specimens have been made. It has been found that grown-in line dislocations are decorated with small (50-100nm diam.) precipitates in all these specimens. Energy dispersive X-ray analysis and spark source mass spectrometry data have been employed in attempts to identify the precipitates.

Reprinted with permission from J.Microscopy, 1980, 118:111 published by Blackwell Scientific Publications Ltd., Oxford, UK.

Aylett, M.R., and Haigh, J.

THE SPECTROSCOPY AND PHOTOLYSIS OF METALLO-ORGANIC PRECURSORS TO
III-V COMPOUNDS

This paper describes the deposition by light-stimulated
breakdown of gallium and indium on borosilicate glass, gallium
arsenide and indium phosphide substrates, from precursors whose
ultraviolet (UV) spectra have recently been reported. Some
preliminary work on the deposition of indium phosphide is also
reported, using an indium-phosphorus complex whose UV spectrum is
also given.

Laser Processing and Diagnostics, Ed. D.Bauerle, Springer Series in
Chem.Phys. 39:263, Springer Verlag, 1984.

Bachrach, R.Z., Bauer, R.S., Chiaradia, P. and Hansson, G.V.

RECONSTRUCTIONS OF GaAs AND AlAs SURFACES AS A FUNCTION OF METAL TO
As RATIO

The room temperature surface reconstructions of GaAs and AlAs
have been investigated with angle integrated photoemission as a
function of the metal to arsenic ratio. Detailed information on
both the (110) and (100) surface as a function of reconstruction has
been obtained from changes in the surface core level intensities,
binding energy shifts, and changes in the surface valence band
density of states. The GaAs(100) surface shows ordered
reconstructions over a wide composition range. The ordering
proceeds through a series of centered structures formed from As and
vacancies in the outer layer in the As rich range to noncentered
structures derived from As, Ga antisite defects, and vacancies on
the Ga rich end. The AlAs(100) surface is predominantly disordered
and only a 3x2 reconstruction was found in a narrow composition
range. Aspects of the Al-Ga exchange reaction on GaAs(110) are
discussed.

Reprinted with permission from J.Vac.Sci.Technol., 1981, 19:335,
published by the American Institute of Physics, New York. © 1981
American Vacuum Society.

Balluffi, R.W.

HIGH ANGLE GRAIN BOUNDARIES AS SOURCES OR SINKS FOR POINT DEFECTS

A secondary grain boundary dislocation climb model for high
angle grain boundaries as sources/sinks for point defects is
described in the light of recent advances in our knowledge of grain

boundary structure. Experimental results are reviewed and are then compared with the expected behavior of the proposed model. Reasonably good consistency is found at the level of our present understanding of the subject. However, several gaps in our present knowledge still exist and these are identified and discussed briefly.

Reprinted with permission from "Grain Boundary Structure and Kinetics":, American Society for Metals, 1980, p1, 290.

Balluffi, R.W., Brokman, A. and King, A.H.

CSL/DSC LATTICE MODEL FOR GENERAL CRYSTAL-CRYSTAL BOUNDARIES AND THEIR LINE DEFECTS

The general CSL/DSC Lattice model for internal boundaries in crystalline materials is described. The model is essentially a 'fit-misfit' model in which the regions of 'fit' are patches where partial lattice matching across the boundary is achieved and the regions of 'misfit' are boundary line defects possessing dislocation/boundary step character. The degree of 'fit' is effectively measured by the size of an appropriate chosen Coincidence Site Lattice (CSL) formed by the two lattices adjoining the boundary. The Burgers vectors of the line defects are vectors of an appropriately chosen DSC Lattice also formed by the two lattices adjoining the boundary, while the step vectors describing the step character are defined within the framework of the DSC Lattice. The model is extremely general and may be applied to grain boundaries in both cubic and non-cubic materials and interphase boundaries. Applications of the model to a variety of experimental results involving grain boundaries and interphase boundaries are discussed. Some general assessment of the current status of the model is attempted.

Reprinted with permission from Acta Met. 1982, <u>30</u>:1453, published by Pergamon Press, Elmsford, New York, USA.

Ban, V.S.

TRANSPORT PHENOMENA MEASUREMENTS IN EPITAXIAL REACTORS

The deposition rates and uniformity in CVD reactors are functions of transport phenomena. It is necessary to understand these phenomena as completely as possible in order to design and use reactors properly. In the last few years several, mostly theoretical, discussions of transport phenomena in CVD reactors appeared. In the present study we discuss the results of the experimental measurements of these phenomena by means of flow visualisation techniques and temperature and concentration

measurements. These measurements suggest a model of flow somewhat
different than models suggested by other authors. The importance of
entry effects and influences on the development of velocity and
temperature profiles are illustrated. Such a model is more complex
than models where the fully developed velocity and/or temperature
fields are assumed, but it is also more descriptive of the true
situation in the reactor.

J.Electrochem.Soc. (1978), $\underline{125}$:317. Reprinted by permission of the
publisher, The Electrochemical Society Inc., Pennington, NJ. USA.

Bardsley, W., Green, G.W., Holliday, C.H., Hurle, D.T.J., Joyce,
G.C., MacEwan, W.R., and Tufton, P.J.

AUTOMATED CZOCHRALSKI GROWTH OF III-V COMPOUNDS

     The weighing method of automatic diameter control has been
extended to grow bulk crystals of gallium arsenide, indium phosphide
and gallium phosphide. An outline of the control theory is given,
and the apparatus and servo loop described.

Reprinted with permission from "GaAs and Related Compounds", 1974.
(Proc.Vth Int.Symp., Deauville). Institute of Physics Conf.Ser.
1975, $\underline{24}$:355.

Barraclough, K.G.

CONTROL OF FOREIGN POINT DEFECTS DURING CZOCHRALSKI SILICON CRYSTAL
GROWTH

     Studies of the axial and radial distribution of interstitial
oxygen, substitutional carbon and metallic contaminants (Fe, Cr) in
82mm diameter Czochralski silicon crystals pulled from melts of well
characterised starting material are reported. Deviations from
normal freeze behaviour are observed and used as a basis to explain
the impurity incorporation mechanisms in terms of some simple
phenomenological models.

Reprinted with permission from "Aggregation Phenomena of Point
Defects in Silicon", Eds. E.Sirtl and J.Goorison, published by the
Electrochemical Society, Pennington, NJ, USA.

Barraclough, K.G., and Series, R.W.

OXYGEN CONTENT OF N+ AND P+ CZOCHRALSKI SILICON

     The axial oxygen contents of antimony- and boron-doped

Czochralski silicon crystals with carrier concentrations between 10^{16} and 10^{19} at. cm^{-3} are compared with the oxygen distributions in lightly-doped crystals ($<10^{15}$ at. cm^{-3}) grown under the same conditions, using calibrated gamma activation and IR absorption analyses. The observed reduction in oxygen content of the antimony-doped crystals at dopant concentrations $> 10^{18}$ at. cm^{-3} is discussed in relation to the silica crucible dissolution rate and the rate of oxygen loss from the melt surface. Studies of the erosion behaviour of silica rods in antimony-doped melts suggest that the addition of antimony to the melt enhances silicon monoxide evaporation through modifications to the surface energy driven flows in the near surface layers.

Reprinted with permission from 'Proc.Symp. on Low-temperature Processing' (Fall Meeting), published by the Electrochemical Society, Pennington, NJ, USA, 1985.

Barnett, S.J., Tanner, B.K., and Brown, G.T.

INVESTIGATION OF THE HOMOGENEITY AND DEFECT STRUCTURE IN SEMI-INSULATING LEC GaAs SINGLE CRYSTALS BY SYNCHROTRON RADIATION DOUBLE CRYSTAL X-RAY TOPOGRAPHY

The high intensity and large beam size of a synchrotron radiation source have been exploited in order to obtain double crystal X-ray topographs of whole 2in. and 3in. slices of semi-insulating LEC GaAs single crystals. Exposure times, typically 30 minutes for high resolution topographs, are at least one order of magnitude down on those required when using a conventional source. Variations in relative lattice parameter and lattice tilt have been measured as a function of position on the slice. The defect structure has been imaged and dislocations are seen in cellular configurations, slip bands and linear arrays (lineage), the latter of which are shown to be associated with small lattice tilts, typically 30". The defect structure revealed on the topographs has been correlated with $1\mu m$ infrared absorption micrographs which are believed to represent the concentration of the dominant deep level EL2.

Reprinted by permission of the publisher from 'Advanced Photon and Particle Techniques for the Characterisation of Defects in Solids', J.B.Roberto, R.W.Carpenter, M.C.Wittels, Eds. Mat.Res.Soc.Symp.Proc. 41:83. Copyright 1985, Elsevier Science Publishing Co.Inc.

Basson, J.H., and Booyens, H.

THE INTRODUCTION OF MISFIT DISLOCATIONS IN HgCdTe EPITAXIAL LAYERS

The variation of misfit with composition and temperature in the HgCdTe/CdTe bicrystal system and its effect on the introduction of misfit dislocations are investigated. It is shown that the critical thickness for the introduction of misfit dislocations is of the order of $0.1\mu m$ for Hg-rich material. Alternative substrate materials to CdTe are considered and the effect of interdiffusion between the substrate and epilayer material on the introduction of dislocations is discussed.

Reprinted with permission from Phys.Stat.Sol.(a) 1983, 80;663.

Bauerle, D.

PRODUCTION OF MICROSTRUCTURES BY LASER PYROLYSIS

Laser pyrolysis at short wavelengths is a powerful tool for micron-sized one-step local deposition of insulating, semiconducting and metallic materials from the gas phase. Various structures, eg stripes on different substrates, and rods of various lengths and diameters, have been produced. The morphology of the deposited material, the deposition rate, and the dimensions (typically $1-300\mu m$) of the structures were investigated quantitatively as functions of laser irradiance, local temperature, laser focus diameter, scanning velocity and gas pressure.

Reprinted by permission of the publisher from 'Materials Research Symposium' Proceedings, 17:19. Copyright, 1983, Elsevier Science Publishing Co.Inc.

Bean, R.C., Zanio, K.R., Hay, K.A., Wright, J.M., Saller, E.J., Fischer, R. and Morkoc, H.

EPITAXIAL CdTe FILMS ON GaAs/Si AND GaAs SUBSTRATES

Epitaxial films of GaAs on (100) silicon were used as substrates for the preparation of high quality epitaxial layers of (100)CdTe, using a simple ultrahigh vacuum technique. Conditions were defined for the consistent formation of the (100)CdTe on (100)GaAs or GaAs/Si. Double crystal X-ray diffraction rocking curves for the GaAs films on Si have shown full width at half maximum as low as 20 s of arc. Both the CdTe and GaAs films are structurally stable to multiple thermal shocks between 80 K and room temperature, and to elevated temperatures in the range of 400 to 500°C.

Reprinted with permission from J.Vac.Sci.Technol. 1986, A4(4):2153, published by the American Institute of Physics. © 1986, American Vacuum Society.

Bell, S.L., and Sen, S.

CRYSTAL GROWTH OF $Cd_{1-x}Zn_xTe$ AND ITS USE AS A SUPERIOR SUBSTRATE FOR
LPE GROWTH OF $Hg_{0.8}Cd_{0.2}Te$

$Cd_{1-x}Zn_xTe$ (x=0.04) boules providing wafers with single crystal
areas as large as 10 to $12cm^2$ have been grown by the vertical
modified-Bridgman (VMB) technique and evaluated as an alternative to
CdTe for use as substrates for LPE growth of HgCdTe layers. The
CdZnTe crystals, as compared with typical CdTe crystals, exhibit
lower defect densities, increased mechanical strength and
significantly improved macro- and micromorphologies of LPE layers of
HgCdTe grown on them. The surface morphology of LPE layers grown on
CdZnTe substrates shows less orientation dependence than for layers
grown on CdTe substrates, for orientations close to the {111}
planes. The addition of Zn to the CdTe lattice may result in
increased covalency and reduced ionicity, which in turn inhibit
plastic deformation and generation of dislocations. These factors,
combined with the ability to adjust the lattice constant to any
desired value within the two extrema, allow the growth of low-defect
density HgCdTe layers required for high-performance IR detector
arrays. The quality of the substrates and epitaxial layers has been
evaluated by defect etching, infrared microscopy, X-ray rocking
curve analysis, and X-ray topography.

Reprinted with permission from J.Vac.Sci.Technol. 1985 A3(1):112,
published by the American Institute of Physics. © 1985, American
Vacuum Society.

Beneking, H., Escobosa, A., and Krautle, H.

CHARACTERISATION OF GaAs EPITAXIAL LAYERS GROWN IN A RADIATION
HEATED MOCVD REACTOR

A new system to grow single crystal epilayers by metal organic
chemical vapor deposition is presented. The graphite susceptor is
heated by a 750W quartz-halogen lamp. To focus the light onto the
backside of the susceptor an elliptical mirror is used. With this
system epilayers of good quality on GaAs are grown down to 600°C.
The morphology, background doping and mobility as a function of
growth conditions is shown. Highly doped layers are grown with H_2Se
and DEZn (diethylzinc).

Reprinted with permission from J.Electron.Mater. 1981, 10(3):473,
published by The Metallurgical Society, Warrendale, PA, USA.

Bensahel, D., Magnea, N., Pautrat, J.L., Pfister, J.C. and Revoil, L.

EFFECT OF ANNEALING AND QUENCHING ON THE NATURE OF THE DOMINANT
ACCEPTORS IN ZnTe

Annealing and quenching studies on high-purity ZnTe have shown
the dominant role of impurities and their solubilities as functions
of stoichiometric conditions. We report also the role of the 'b'
acceptor related to lithium, the 'g' acceptor related to silver and
the 'a' acceptor of unknown nature.

Reprinted with permission from 'Defects and Radiation Effects in
Semiconductors', Proc.Int.Conf., Nice, Sept 1978.
Inst.Phys.Conf.Ser. 46:421, published by the Institute of Physics,
Bristol, UK.

Bourret, A., Thibault-Desseaux, J. and Seidman, D.N.

EARLY STAGES OF OXYGEN SEGREGATION AND PRECIPITATION IN SILICON

The early stages of oxygen segregation at dislocations and
precipitation in the bulk have been investigated by high-resolution
electron microscopy in Czochralski-grown silicon. Two kinds of
precipitates are observed: a crystalline silica phase, coesite and
an amorphous phase. Both forms coexist after 650°C heat treatment:
the so-called rodlike defects are in fact long <011> ribbons of
coesite associated with interstitial dipoles. This crystalline form
is favored by a high oxygen supersaturation and a low carbon
content. Above 870°C amorphous platelets of silica are formed on
the (100) planes, whereas coesite is no longer observed but
interstitial dislocation loops are always present. The strain
produced by such precipitates is partially relaxed by Si
interstitial emission, which explains the internal formation of
dislocations. It is suggested that both forms are nucleated on two
different species of nuclei. At <011> dislocation cores it is shown
that the coesite phase is stablised over a wide range of oxygen
content or annealing temperature.

Reprinted with permission from J.Appl.Phys. 1984, 55(4):825,
published by the American Institute of Physics, New York.

Boyd, I.W.

LASER ASSISTED PYROLYTIC GROWTH AND PHOTOCHEMICAL DEPOSITION OF THIN
OXIDE FILMS

Lasers offer a diverse approach to fabricating oxide films for
a wide variety of applications. This paper will review the field of
laser oxidation in general but particular attention will be focused

on oxide formation on silicon. Two main areas will be detailed, namely Laser Chemical Vapour Deposition (LCVD) and Laser Pyrolytic Growth (LPG). The kinetics involved with film preparation will be discussed while the various structural, optical and electrical properties of the films will be compared.

Reprinted with permission from "Laser Processing and Diagnostics", Ed. D.Bauerle, published by Springer Verlag, Berlin, 1984.

Bozler, C.O.

REDUCTION OF AUTODOPING

Silicon epitaxial layers are grown on silicon substrates highly doped with arsenic using the pyrolysis of silane. A special technique is used to reduce the autodoping by minimising impurity evaporation from the back side of the wafer. A comparison of impurity profiles obtained from layers grown under a number of different conditions reveals that autodoping can be reduced still further by increasing the flow rate, decreasing the volume around the susceptor, and decreasing the growth rate. The formulation of a mathematical model helps to describe the interrelation of those three variables. By optimising the growth conditions, it is shown that the autodoping can be essentially eliminated.

J.Electrochem.Soc. (1975) 122(12):1705. Reprinted by permission of the publisher, The Electrochemical Society Inc., Pennington, NJ, USA.

Brown, G.T., Cockayne, B., and MacEwan, W.R.

THE NATURE OF PRISMATIC DISLOCATION LOOPS IN UNDOPED InP

The rectangular prismatic dislocation loops found in single crystals of InP grown from indium-rich melts have been analysed further by transmission electron microscopy (TEM) and a model is proposed for their formation. These loops are shown to be of an interstitial nature and it is proposed that they form by condensation of indium interstitials onto dislocation lines causing climb. Because this climb occurs in the presence of only one native interstitial species, the climbed plane is expected to be imperfect, consisting of either V_P or In_P. Such a defective plane will create a dilation parallel to the plane normal and lead to the displacement fringes observed by TEM. The defective plane is shown to act as a nucleation site for subsequent precipitation at low temperature. EDX analysis suggests that this precipitate is indium.

Reprinted with permission from J.Electron.Mater. 1983, 12:93, published by the Metallurgical Society of AIME, Warrendale, PA, USA.

Cadoret, R. and Cadoret, M.

A THEORETICAL TREATMENT OF GaAs GROWTH BY VAPOUR PHASE TRANSPORT FOR {001} ORIENTATION

The normal growth rate of a {001} face has been theoretically studied: by considering either direct fixation of gallium arsenide molecules, or formation of intermediate surface compounds. From the theory of rate processes, it appears that experimental results can be interpreted by considering the reactions of desorption of the chlorine atoms adsorbed on surface as limiting the growth. A theoretical expression of the normal growth rate based on desorption by hydrogen has been performed. The descending portions of the curves with decreasing substrate temperature or increasing partial pressure of gallium monochloride appear as due to an increasing coverage of surface with gallium monochloride molecules. Absolute theoretical values agree with experimental published measures, except for the weakest substrate temperatures. This disagreement may be due to the possibility of desorption of two chlorine atoms by gallium monochloride and formation of gallium trichloride molecules.

Reprinted with permission from J.Cryst.Growth, 1975, 31:142, published by North Holland Publishing Co., Amsterdam.

Calawa, A.R.

ON THE USE OF AsH_3 IN THE MOLECULAR BEAM EPITAXIAL GROWTH OF GaAs

High-quality epitaxial layers of GaAs have been grown in a molecular beam epitaxial system using AsH_3 as the arsenic source. Peak electron mobilities of over 130 000 cm^2/V sec and 77-K mobilities as high as 110 000 cm^2/V sec have been observed in a 5 μm-thick GaAs layer with a carrier concentration of 2.4×10^{14} cm^{-3}. These layers are grown on Cr-doped semi-insulating GaAs substrates. Initial results indicate that As_1 may be the preferred specie for the growth of high-purity GaAs.

Reprinted with permission from Appl.Phys.Lett. 1981, 38(9):701, published by the American Institute of Physics, New York.

Capper, P., Gosney, J.J. and Jones, C.L.

APPLICATION OF THE ACCELERATED CRUCIBLE ROTATION TECHNIQUE TO THE BRIDGMAN GROWTH OF $Cd_xHg_{1-x}Te$: SIMULATIONS AND CRYSTAL GROWTH

Simulation studies have been carried out to determine the fluid flows induced in the tall narrow containers used in the Bridgman growth of $Cd_xHg_{1-x}Te$ by the influence of ACRT. Spiral shearing,

Ekman and transient Couette flows have all been observed. Ekman
flow and Couette flows are the dominant stirring mechanisms at the
crystal/melt interface and crucible wall respectively. A comparison
is given between experimentally determined critical rotation rates
for the onset of Couette flow and those predicted using Rayleigh's
criterion. Crystals of $Cd_xHg_{1-x}Te$ grown using ACRT are found to
have a much greater degree of axial and radial composition
uniformity than equivalent non-ACRT crystals. In addition an
improvement in crystallinity has been obtained which is thought to
be due to Couette flows at the crucible walls.

Reprinted with permission from J.Cryst.Growth, 1984, 70:356,
published by North Holland Publishing Co., Amsterdam.

Capper, P., Gosney, J.J., Jones, C.L., Kenworthy, I. and Roberts,
J.A.

ACCEPTOR DOPING OF BRIDGMAN-GROWN $Cd_xHg_{1-x}Te$

    The behaviour of various dopants known to be acceptors in
$Cd_xHg_{1-x}Te$ has been studied in crystals grown by the Bridgman
process. Dopants from Groups I (Li, Cu, Ag) and V (P, As, Sb) were
added, in elemental form, to the initial melts. Chemical analysis
was used to determine the segregation behaviour along the grown
crystals and, linked to Hall effect measurements, to establish the
electrical activity of each dopant. More extensive studies were
then carried out on examples (Ag and Sb) from each group. These two
dopants are believed to be the slowest diffusing species within
their group, making electrical characterisation and doping control
easier. Linear relationships were found between carrier
concentration and amount of dopant added for both elements. Hall
effect measurements down to 20 K have been performed and acceptor
ionisation energies determined. The values for doped material are
compared to those from samples containing native acceptor defects
and are found to be shallower. Isothermal annealing in mercury
vapour has been used to study the stability of the dopants and to
assess the suitability of the material for use as substrates for
diode fabrication by ion implantation.

Reprinted with permission from J.Cryst.Growth, 1985, 71:57,
published by North Holland Publishing Co., Amsterdam.

Carruthers, J.R.

ORIGINS OF CONVECTIVE TEMPERATURE OSCILLATIONS IN CRYSTAL GROWTH
MELTS

    Sinusoidal temperature oscillations of large amplitude and with
periods ranging from a few seconds to a few minutes have been

reported for many different crystal growth melt materials and configurations. In this paper, several models of the origins of these temperature oscillations are described which successfully explain the results reported for various materials. At low values of the Rayleigh number, the instability of the convection roll itself in the form of transverse waves is the primary source of temperature oscillations. At higher values of the Rayleigh number, periodic boundary layer instabilities develop in thermal layers for high Prandtl number fluids (such as molten oxides) and momentum layers for low Prandtl number fluids (such as molten metals and semiconductors).

Reprinted with permission from J.Cryst.Growth, 1976, $\underline{32}$:13, published by North Holland Publishing Co., Amsterdam.

Chang, C.E., Wilcox, R.

CONTROL OF INTERFACE SHAPE IN THE VERTICAL BRIDGMAN-STOCKBARGER TECHNIQUE

At very slow travel rates the solid-liquid interface shape (from the viewpoint of the solid) is convex when it lies within the heater and concave when it lies within the cooler. The dependence of interface position on travel rate, geometry of the ampoule, and on the relative temperatures of the heater and cooler increases as the effectiveness of heat transfer between ampoule and the surroundings diminishes and as the thermal conductivity of the material increases (at least for low travel rates).

Reprinted with permission from J.Cryst.Growth, 1974, $\underline{21}$:135, published by North Holland Publishing Co., Amsterdam.

Chang, H.R.

AUTODOPING IN SILICON EPITAXY

Vertical and lateral autodoping in silicon epitaxy deposited in a vertical pancake-type reactor were examined as a function of pre-epitaxial bake cycle, dopant type, source gas, and growth parameters. Boron, phosphorus, arsenic, and antimony were used as dopant elements to form a localised buried layer at the center of substrates. Silane, dichlorosilane, and silicon tetrachloride were used as the silicon source for epitaxial growth. The results show that higher prebaking temperature produces more lateral autodoping for boron and phosphorus, while the opposite holds for arsenic and antimony. Autodoping in both vertical and lateral directions decreases with increasing prebake time. A continual decrease in lateral autodoping is found with increasing growth rate for phosphorus; for arsenic, only a slight difference is noticed in the growth-rate range studied. Phosphorus and arsenic show a similar trend at the deposition temperature of lateral autodoping, ie autodoping decreases at first, then increases with increasing temperature. For all the dopants studied, vertical and lateral autodoping decrease with increasing number of chlorine atoms in the source gases.

J.Electrochem.Soc. (1985), 132(1):219. Reprinted by permission of the publisher, The Electrochemical Society Inc., Pennington, NJ. USA.

Chen, C.J., and Osgood, R.M. Jnr.

SPECTROSCOPY AND PHOTOREACTIONS OF ORGANOMETALLIC MOLECULES ON SURFACES

A study of the UV photochemistry of organometallic molecules in the vapor and adsorbed phases is discussed. The gas-phase spectra is interpreted on the basis of molecular orbital theory and experimental data. The absorption spectra of the surface adlayers are determined by computer subtraction of the gas-phase spectra and an interpretation is provided. The influence of metal-substrate microstructure on UV photochemical reactions of metal alkyls is briefly discussed.

Reprinted by permission of the publisher from "Laser Diagnostics and Photochemical Processing for Semiconductor Devices", Ed. R.M.Osgood, S.R.J.Brueck and H.R.Schlossberg, Mater.Res.Soc.Symp.Proc. 17:169, 1983. Copyright Elsevier Science Publishing Co.Inc.

Chen, K.

HEAT AND MASS TRANSFER IN HORIZONTAL VAPOR PHASE EPITAXY REACTORS

Heat and mass transfer processes were studied in a horizontal vapor epitaxy reactor with simultaneously developing velocity, temperature and concentration distributions. The susceptor placed in the reactor was either horizontal or slightly tilted. An analytical solution was obtained using previous heat transfer results in entry length and fully developed duct flow, and the heat-mass transfer analogy for growth rate calculations. The mean concentration distributions along the susceptor are presented in dimensionless forms based on constant properties and laminar flow. The influences of axial temperature variation and wall cooling on the epitaxial growth rate distribution are examined in this study. Results of experimental studies of silicon growth from SiH_4 in hydrogen agree well with the analysis.

Reprinted with permission from J.Cryst.Growth, 1984, 70(1/2):64,
published by North Holland Publishing Co., Amsterdam.

Chen, R.T. and Holmes, D.E.

DISLOCATION STUDIES IN 3-INCH DIAMETER LIQUID ENCAPSULATED
CZOCHRALSKI GaAs

We have characterised the density and distribution of
dislocations in 3-inch diameter, undoped GaAs crystals grown by the
liquid encapsulated Czochralski technique. The radial distribution
across wafers follows a "W"-shaped profile indicating excessive
thermal gradient-induced stress as the primary cause of
dislocations. The density along the body of each crystal increases
continuously from front to tail. In contrast, the longitudinal
distribution in the cone region is inverted, first increasing, and
then decreasing as the crystal expands from the neck to full
diameter. Growth parameters favoring reduced dislocation densities
include good diameter control, the use of thick B_2O_3 encapsulating
layers, slightly As-rich melts, and low ambient pressures. The
dislocation density in the body of the crystal is practically
independent of cone angle θ for $20°C < \theta < 70°$. However, high
densities result for flat-top ($0° < \theta < 20°$) crystals. Dash-type
seed necking works to reduce the dislocation density only when
high-density seeds (> 5000 cm^{-2}) are used. Dislocation densities
below 1×10^4 cm^{-2} can be achieved routinely in the large annulus
region at the front of a crystal through proper control of these
parameters. The lowest measured dislocation density was 6000 cm^{-2}.
Further, we show that convective heat transfer from the crystal to
the high pressure ambient plays a dominant role in controlling the
dislocation density.

Reprinted with permission from J.Cryst.Growth, 1983, 61:111,
published by North Holland Publishing Co., Amsterdam.

Chew, N.G., Cullis, A.G., Warwick, C.A., Robbins, D.J., Hardeman,
R.W., and Gasson, D.B.

ELECTRON MICROSCOPE CHARACTERISATION OF SILICON LAYERS GROWN BY
MOLECULAR BEAM EPITAXY

Transmission electron microscopy has been used to study
epitaxial Si layers deposited onto single crystal Si substrates by
molecular beam epitaxy. The structural results obtained have been
correlated with spectrally resolved luminescence data acquired using
an SEM based low temperature cathodoluminescence system. The
dependence of layer defect structure upon surface cleaning
procedures and deposition temperature have been characterised and

the effects of particle inclusion during growth determined.

Reprinted with permission from "Electron Microscope Characterisation
of Silicon Layers Grown by Molecular Beam Epitaxy" in 'Microscopy of
Semiconducting Materials', Eds. A.G.Cullis and D.B.Holt., Institute
of Physics Conf.Ser. 1985, 76:129.

Cho, A.Y.

EPITAXIAL GROWTH OF GALLIUM PHOSPHIDE ON CLEAVED AND POLISHED (111)
CALCIUM FLUORIDE

The mechanism of growth of GaP on a cleaved CaF_2 (111) surface
was studied in situ in a high-energy reflection electron diffraction
system. It was found that in the early stages of growth, GaP forms
tetrahedral nuclei with {111} faces. The three edges of the
tetrahedron are parallel to the three ⟨110⟩ directions. These
microcrystals coalesce and form a smooth film after a mean thickness
of more than 300 monolayers of GaP is deposited on the surface.
Temperatures for epitaxial growth of a single crystal without
twinning as a function of the atom arrival rate were studied for GaP
on a clean CaF_2 surface and on a GaP-covered CaF_2 surface. It was
found that growing GaP without twinning on a bare cleaved CaF_2
surface requires a temperature ~65°C higher than on a surface that
is covered with GaP. The structural characteristics of the GaP film
as a function of the substrate temperature are also discussed.

Reprinted with permission from J.Appl.Phys. 1970, 41(2):782
published by The American Institute of Physics, New York.

Cho, A.Y.

FILM DEPOSITION BY MOLECULAR-BEAM TECHNIQUES

A review of molecular-beam epitaxy of GaAs and the observation
of surface structures with high-energy electron diffraction in an
ultrahigh-vacuum system is described. The utilisation of these
surface structures as growth conditions to produce n- and p-type
layers when doped with Sn, Ge, and Mg, and the electrical and
optical evaluations of the layers thus grown is also discussed. The
molecular-beam epitaxy method may be used to fabricate extremely
thin multilayer structures and may play an increasing role in
semiconductor technology.

Reprinted with permission from J.Vac.Sci.Technol. 1971, 8(5):531,
published by the American Institute of Physics. © 1971 American
Vacuum Society

Cho, A.Y., and Dernier, P.D.

SINGLE CRYSTAL ALUMINIUM SCHOTTKY-BARRIER DIODES PREPARED BY
MOLECULAR-BEAM EPITAXY (MBE) ON GaAs

Metal-semiconductor surface barriers were formed by epitaxially
grown Al(001) on GaAs(001) with MBE at room temperature. The diodes
exhibit nearly ideal electrical characteristics. It is also
demonstrated that different barrier heights may be achieved by
growing Al layers on MBE GaAs surfaces with different surface
stoichiometries.

Reprinted with permission from J.Appl.Phys. 1978, 49(6):3328,
published by the American Institute of Physics, New York.

Cho, A.Y.

GROWTH AND PROPERTIES OF III-V SEMICONDUCTORS BY MOLECULAR BEAM
EPITAXY

Advances in solid-state device technology in the sixties
established III-V materials as a new class of semiconductors for
high-speed microwave and highly efficient optical devices.
Molecular beam epitaxy (MBE) is an extremely versatile thin film
technique which can produce single crystal layers with atomic
dimensional controls and thus permit the preparation of novel
structures and devices tailored to meet specific needs. Important
factors to achieve high quality MBE growth such as *in situ* analysis,
substrate preparations, growth conditions and layer properties are
discussed.

Reprinted with permission from "Growth and properties of III-V
semiconductors by molecular beam epitaxy", p191 in "Molecular Beam
Epitaxy and Heterostructures", Eds. L.L.Chang and K.Ploog, published
by Martinus Nijhoff, Dordrecht, The Netherlands, 1985.

Chow, R., and Chai, Y.G.

A PH₃ CRACKING FURNACE FOR MOLECULAR BEAM EPITAXY

A furnace was fabricated to crack phosphine in ultrahigh
vacuum. Thermodynamic calculations showed that phosphine should
dissociate into the phosphorus dimer to a high degree. Furnace
materials were compared by construction of all-tantalum and
all-quartz cartridges which slid into the furnace hot zone.
Phosphine was found to react with tantalum, but not to any great
extent with quartz. The effects of baffles, studied with the quartz

cartridges, lowered the temperature at which the cracking
efficiencies began to increase and increased cracking efficiencies.
The temperature range studied was from 100° to 1170°C.

Reprinted with permission from J.Vac.Sci.Technol. 1983, A1(1):49,
published by the American Institute of Physics. © 1983 American
Vacuum Society.

Claeys, C., Bender, H., Declerck, G., Van Landuyt, J., Van
Overstraeten, R., and Amelinckx, S.

KINETICS OF PROCESS-INDUCED DEFECT COMPLEXES IN SILICON

High temperature processing of Czochralski silicon material
introduces a large variety of defect complexes. High voltage
electron microscopy is used to identify the structure of the defects
observed in samples annealed in respectively inert and oxidizing
ambients over a wide temperature and time range. The influence of
the atmosphere on the denuded zone formation in the near-surface
region of the wafers is discussed.

Special attention is given to the nucleation and growth
kinetics of the stacking faults. A mechanism to explain the
transformation from multiple Frank dislocation loops, which nucleate
by a Bardeen-Herring process, into 1/6 <114> dislocation loops is
proposed. Experimental evidence to demonstrate the validity of this
model will be given. The behaviour of prismatic punching systems is
analysed as a function of anneal temperature.

Reprinted with permission from "Aggregation Phenomena of Point
Defects in Silicon", Eds. E.Sirtl and J.Goorison, published by the
Electrochemical Society, Pennington, NJ, USA, 1982.

Cockayne, B., and Roslington, J.M.

THE DISLOCATION-FREE GROWTH OF GADOLINIUM GALLIUM GARNET SINGLE
CRYSTALS

It is established that a substantial number of the microscopic
strain centres observed in Czochralski-grown gadolinium gallium
garnet single crystals are due to dislocations whose propagation is
markedly dependent upon the shape of the solid/liquid interface. It
is shown that the dislocation density can be minimised by using
facets to block dislocation propagation and that crystals
substantially free from dislocations can be produced.

Reprinted with permission from J.Mater.Sci., 1973, 8:601, published
by Chapman and Hall Ltd., London.

Cockayne, B., Brown, G.T., and MacEwan, W.R.

DISLOCATION CLUSTERS IN CZOCHRALSKI-GROWN SINGLE CRYSTAL INDIUM PHOSPHIDE

Clusters of etch pits which sometimes form in single crystal indium phosphide grown by the Czochralski liquid-encapsulation technique, are shown to equate with prismatic dislocation loops generated by a stress source at the centre of the cluster. The morphology of the clusters and the conditions under which such clusters form are shown to be consistent with the presence of either a precipitate or an inclusion. It is demonstrated that water vapour associated with the boric oxide used as the liquid encapsulant is a positive cause of clusters and that vacuum baking is a pre-requisite for their control. The elements responsible for the stress generation promoting the clusters have not been positively identified but the likely possibilities are discussed.

Reprinted with permission from J.Cryst.Growth, 1981, 51:461, published by North Holland Publishing Co., Amsterdam.

Cockayne, B., Brown, G.T. amd MacEwan, W.R.

CONTROL OF DISLOCATION STRUCTURES IN LEC SINGLE CRYSTAL InP

It is shown here that formation of the major dislocation structures detected in single crystal InP, such as clusters and large rectangular prismatic loops, can be prevented during LEC growth. The conditions under which this control can be applied are entirely compatible with the techniques used to inhibit dislocation formation both by slip and the propagation of dislocations from the seed. In consequence, InP crystals with very low densities can now be produced.

Reprinted with permission from J.Cryst.Growth, 1983, 64:48, published by North Holland Publishing Co., Amsterdam.

Cole, S.

PLASTIC BENDING OF $Cd_xHg_{1-x}Te$

Preliminary results are reported of three-point plastic bending tests on $Cd_xHg_{1-x}Te$ single crystal samples, for an x value of about 0.2, conducted in air at strain rates of the order of $10^{-5}sec^{-1}$, and at temperatures in the range 303K (30°C) to 363K (90°C) (in the region of 0.35 T_M, where T_M is the absolute melting point). Single crystal samples were cut from polycrystalline ingots, and the orientation, although measured in each case, was not consistent from

sample to sample, being determined by the available grain shape.
The stress-strain curves resemble those found for Group IV and III-V
semiconductors. They display a yield drop, followed by a region of
zero work hardening. All tests were stopped in this region, and in
no case did the overall glide strain exceed 3%. The upper and lower
yield stresses (outer fibre glide stress values) varied from 16MN
m^{-2} and 10MN m^{-2}, respectively, at 363K (90°C) to 24MN m^{-2} and 17MN
m^{-2}, respectively, at 303K (30°C).

Reprinted with permission from J.Mater.Sci., 1980, 15:2591,
published by Chapman and Hall Ltd., London.

Coltrin, R.J., Kee, R.J., and Miller, J.A.

A MATHEMATICAL MODEL OF THE COUPLED FLUID MECHANICS AND CHEMICAL
KINETICS IN A CHEMICAL VAPOR DEPOSITION REACTOR

We describe a numerical model of the coupled gas-phase
hydrodynamics and chemical kinetics in a silicon chemical vapour
deposition (CVD) reactor. The model, which includes a 20-step
elementary reaction mechanism for the thermal decomposition of
silane, predicts gas-phase temperature, velocity, and chemical
species concentration profiles. It also predicts silicon deposition
rates at the heated reactor wall as a function of susceptor
temperature, carrier gas, pressure and flow velocity. We find
excellent agreement with experimental deposition rates, with no
adjustment of parameters. The model indicates that gas-phase
chemical kinetic processes are important in describing silicon CVD.

J.Electrochem.Soc. (1984), 131:425. Reprinted by permission of the
publisher, The Electrochemical Society Inc., Pennington, NJ, USA.

Crochet, M.J., Wouters, P.J., Geyling, F.T., and Jordan, A.S.

FINITE-ELEMENT SIMULATION OF CZOCHRALSKI BULK FLOW

Steady state and time-dependent finite element simulations have
been generated to represent transport phenomena in a Czochralski
melt. Three basic convection mechanisms, previously recognised in
the literature, have been verified. Their combined effects have
been explored and their extension into the oscillatory regime has
been demonstrated. The transient code will permit the modeling of
interactions between the melt and transport phenomena in the crystal
across the time-variable interface during actual crystal growth.

Reprinted with permission from J.Cryst.Growth, 1983, 65:153,
published by North Holland Publishing Co., Amsterdam.

Crochet, M.J., Geyling, F.T. and Van Schaftingen, J.J.

NUMERICAL SIMULATION OF THE HORIZONTAL BRIDGMAN GROWTH OF A GALLIUM
ARSENIDE CRYSTAL

Two-dimensional numerical simulations have been developed which
represent the thermomechanical behavior of semiconductor melts in
horizontal crucibles. These computer models are based on
time-dependent finite-difference and finite-element codes, capable
of simulating steady and transient melt convection. At moderate
Rayleigh numbers, one finds time-invariant solutions which typically
involve several primary and secondary vortices, depending on aspect
ratio. Above a critical Rayleigh number, these steady solutions are
replaced by oscillatory melt convection, as suggested by earlier
experimental and anlytic studies. The finite element code was
extended to include a solid-liquid interface, which permits the
study of actual crystal growth processes. In the presence of
thermal oscillations, one finds periodic freezing and remelting at
the interface, which may be related to the observed striations in
crystals grown by the gradient freeze technique. The amplitude of
these oscillations depends on the boundary conditions and the Stefan
number. Three-dimensional extensions of the steady-state code are
under way to determine the effects of finite crucible width, as
encountered in experiments.

Reprinted with permission from J.Cryst.Growth, 1983, 65:166,
published by North Holland Publishing Co., Amsterdam.

Cullen, G.W., Duffy, M.T., Jastrzebski, L., and Lagowski, J.

THE CHARACTERISATION OF CVD SINGLE-CRYSTAL SILICON ON INSULATORS:
HETEROEPITAXY AND EPITAXIAL LATERAL OVERGROWTH

In the development of the heteroepitaxial silicon technology,
progress had been impeded because of the labor and time involved in
the evaluation of the crystalline quality of the deposits by
transmission electron microscopy and device performance. Therefore
an important part of our program has been directed toward the
development of rapid and non-destructive methods for the
characterization of the silicon crystallinity. It is essential that
the results of such methods relate to device performance. In this
paper, we update earlier reports on this effort. Emphasis has been
placed on the use of UV reflectometry for probing the near-surface
silicon quality, photovoltage spectroscopy for probing the "bulk" of
the film, interference photovoltage spectroscopy for probing the
near-substrate portion of the silicon, and infrared multiple
reflectometry to assess the crystalline quality of the substrate
surface. The silicon-on-insulator effort has recently been extended
on the growth of dielectrically isolated silicon by CVD epitaxial

lateral overgrowth. The evaluation of these deposits by
transmission electron microscopy is described.

Reprinted with permission from J.Cryst.Growth, 1983, 65:415,
published by North Holland Publishing Co., Amsterdam.

Cullis, A.G., and Katz, L.E.

ELECTRON MICROSCOPE STUDY OF ELECTRICALLY ACTIVE IMPURITY
PRECIPITATE DEFECTS IN SILICON

A specific defect which was found to electrically degrade
junctions in Si transistor devices has been studied using both
transmission and scanning electron microscopy, together with
electron probe X-ray micro-analysis. It was observed that the
defect had a characteristic rod-like morphology, penetrating the
[001] Si foils along inclined $\langle 101 \rangle$ directions. Both electron
diffraction and X-ray micro-analytical data indicated that the rods
had a precipitate structure consistent with that of $\int_a$-FeSi$_2$
(α-lebolite). The rod/matrix orientation relationship was found to
be (110)FeSi$_2$ $\|$(101)Si and (111) FeSi$_2$~$\|$(010)Si, each rod axis being
a [110]FeSi$_2$ direction. Furthermore, rod defects often occurred
with accompanying small subsidiary precipitates thought to be a Cu
silicide, which either decorated the rods themselves or formed
planar colonies enclosed within irregular dislocation loops attached
to the rods.

The rod precipitate growth is discussed both with reference to
the orientation relationship with the matrix and also with regard to
the diffusion characteristics of Fe in Si. A kinetic model is
constructed to provide a comparison with Cu silicide precipitate
growth.

The presence of $\int_a$-FeSi$_2$ precipitate rods in device structures
is most undesirable since this phase exhibits metallic conduction
properties. Because it is necessary to remove rod defects from
transistor devices in order to restore satisfactory electrical
properties, the use of the P-diffusion gettering technique is also
examined and shown to be effective for this purpose.

Reprinted with permission from Phil.Mag., 1974, 30(6):1419,
published by Taylor & Francis Ltd., London, UK.

Cullis, A.G., Augustus, P.D., and Stirland, D.J.

ARSENIC PRECIPITATION AT DISLOCATIONS IN GaAs SUBSTRATE MATERIAL

High-resolution diffraction utilizing an analytical electron

microscope has been employed to identify small (~500A-diam)
precipitates attached to line dislocations in gallium arsenide. The
results show that the precipitates consist of crystallites of
elemental hexagonal arsenic embedded within the gallium arsenide
matrix. Precipitates were observed in a range of semi-insulating,
p-type, and n-type material and were not dependent on the presence
of specific additional dopants for their occurrence. The way in
which the particles may originate is discussed.

Reprinted with permission from J.Appl.Phys. 1980, 51(5):2556,
published by the American Institute of Physics, New York.

Curtis, B.J., and Dismukes, J.P.

EFFECTS OF NATURAL AND FORCED CONVECTION IN VAPOR PHASE GROWTH
SYSTEMS

Convective temperature oscillations (CTO) introduce growth
striae and other inhomogeneities in melt grown crystals, whereas a
literature review suggests that the cause of such defects in
materials grown from the vapor phase by heterogeneous chemical
reaction, either in closed tubes or in flow-through systems, is not
satisfactorily understood. Although CTO occur in room temperature
gases at values of Rayleigh number (R) greater than about 6×10^3, it
was not clear whether CTO occur in practical open tube chemical
vapor deposition systems at elevated temperature, because numerical
calculations indicated that at constant pressure the R of gases
depends on temperature as T^{-n}, with $n \approx 4.5$.

Measurements carried out in the present work in flowing systems
similar to those used in open tube chemical vapor deposition have
shown that, in general agreement with hydrodynamic predictions, CTO
with amplitude up to 20C and frequency up to 0.25 cycles/sec occur
both in vertical and horizontal apparatus with cold walls, but not
in apparatus with hot walls. Hydrogen and helium are found
experimentally to be very much more stable than argon or nitrogen
under equivalent conditions, also in agreement with Rayleigh number
calculations. The observed superiority from a hydrodynamic
stability viewpoint of hydrogen compared to helium, is an important
anomaly, since Rayleigh number calculations indicate helium should
be slightly more stable than hydrogen over the range from room
temperature to 1200K.

Reprinted with permission from J.Cryst.Growth, 1972, 17:128,
published by North Holland Publishing Co., Amsterdam.

Dobson, P.S., Fewster, P.F., Hurle, D.T.J., Hutchinson, P.W.,
Mullin, J.B., Straughan, B.W. and Willoughby, A.F.W.

SUPERDILATION AND DEFECTS IN TELLURIUM-DOPED GALLIUM ARSENIDE

    Precision lattice parameter measurements on Te-doped GaAs, with
free-electron concentration 4-8 x 10^{18} cm^{-3}, confirmed
'superdilation', that is, a much greater expansion of the lattice
than simple theory predicts. Annealing in vacuo at 880°C for 4 h
produced a radical decrease in parameter, growth of interstitial
loops, and changes in electrical activity, all of which are
quantitatively connected to the original free-electron concentration. Possible defect models are discussed.

Reprinted with permission from "Superdilation and defects in
tellurium-doped gallium arsenide' in "GaAs and Related Compounds".
Inst. Physics Conf.Series, 1979, 45:16.

Dobson, P.J., Neave, J.H., and Joyce, B.A.

RHEED EVIDENCE FOR A DOMAIN STRUCTURE OF GaAs(001)-2x4 AND -4x2
RECONSTRUCTED SURFACES

Curved streaks in RHEED patterns from MBE-prepared reconstructed
(001) GaAs surfaces are shown to originate from antiphase domains
formed by tilted As-As dimer chains. Several possible domain
boundary configurations are proposed, including both coplanar and
stepped structures.

Reprinted with permission from Surf.Sci. 1982, 119:L339, published
by North Holland Publishing Co., Amsterdam.

Donahue, T.J., Burger, W.R. and Reif, R.

LOW-TEMPERATURE SILICON EPITAXY USING LOW PRESSURE CHEMICAL VAPOR
DEPOSITION WITH AND WITHOUT PLASMA ENHANCEMENT

Specular epitaxial silicon films have been deposited at 775°C using
a low pressure chemical vapor deposition process both with and
without plasma enhancement. This is the lowest silicon epitaxial
deposition temperature reported for thermally driven chemical vapor
deposition. It was found that the predeposition in situ cleaning of
the surface, rather than any plasma effects during the deposition,
was essential for achieving epitaxial growth at this low
temperature. Surface cleaning in these experiments was done by
sputtering the wafer in an argon plasma at 775°C with a dc bias
applied to the susceptor. This is the lowest pre-epitaxial cleaning
temperature reported for thermally driven chemical vapor deposition.

Reprinted with permission from Appl.Phys.Lett. 1984, 44(3):346,
published by the American Institute of Physics, New York.

Donnelly, V.M., Geva, M., Long, J., and Karlicek, R.F.

EXCIMER LASER INDUCED DEPOSITION OF InP AND INDIUM-OXIDE FILMS

InP and In-oxide films have been deposited on quartz, GaAs and
InP substrates by excimer laser induced photodecomposition of
$(CH_3)InP(CH_3)_3$ and $P(CH_3)_3$ vapors at 193nm. The oxide film
refractive index and stoichiometry are close to In_2O_3. Phosphorus
incorporation in the films was greatly enhanced by focusing the
laser beam to promote multiple-photon dissociation processes. These
conditions also lead to enhanced carbon inclusion in the films, due
to formation of species such as CH and CH_2 in the gas phase.
However, this carbon inclusion could be suppressed by focusing the
beam onto the surface at normal incidence. In the irradiated zone
InP could be deposited with $P(CH_3)_3$-to-$(CH_3)_3InP(CH_3)_3$ ratios of
only ~1:1. The technique offers several potential advantages over
conventional metalorganic chemical vapor deposition, including lower
temperature, enhanced rates, safer gases, and three-dimensional film
composition control. Strong atomic In emission is observed in the
gas phase above the depositing film, due to a multiple photon
dissociation process. Gas phase fluorescence from P, CH and C was
also observed. These emissions give insight into the
photodecomposition mechanism and also serve as a monitor of
metalorganic precursor concentrations.

Reprinted with permission from Appl.Phys.Lett. 1984, 44(10):951,
published by the American Institute of Physics, New York.

Duchemin, J-P., Bonnet, M. and Koelsch, F.

KINETICS OF SILICON GROWTH UNDER LOW HYDROGEN PRESSURE

The kinetics of silicon deposition have been studied in a
reactor working under reduced hydrogen pressure between 10 Torr and
atmospheric pressure (760 Torr) for the following silicon sources:
SiH_4, SiH_2Cl_2 and $SiCl_4$. In every case, the kinetics are controlled
by the surface at low hydrogen pressure whereas at higher pressure
the mass transfer becomes slower. When the kinetics are controlled
by the surface, the deposition rate is inversely proportional to the
square root of the hydrogen pressure and is activated. In the
opposite case the deposition rate is inversely proportional to the
hydrogen pressure and does not depend on the temperature. This
result, and the fact that the depositions are monocrystalline even
at low temperature when the pressure is lower than 70 Torr, show

that the hydrogen is strongly adsorbed on the silicon surface. The
deposition made at low pressure and low temperature have abrupt
impurity profiles which allow us to realise good quality microwave
components.

J.Electrochem.Soc. (1978), 125:651. Reprinted by permission of the
publisher, The Electrochemical Society Inc., Pennington, NJ, USA.

Duchemin, J.P.

LOW PRESSURE CVD OF III-V COMPOUNDS

In this paper it is shown that the low pressure organometallic
chemical vapor deposition (LP-MOCVD) technique of growing
semiconductors is generally applicable to most of the III-V
compounds that are currently of interest. The principles of the
technique are described and then specific applications are
detailed. It has been found that the LP-MOCVD technique offers the
following advantages: reduction of autodoping, virtual elimination
of parasitic reactions and the possibility of growth on large areas
of semiconductor substrates.

Reprinted with permission from J.Vac.Sci.Technol., 1981, 18(3):753,
published by the American Institute of Physics © 1981, American
Vacuum Society.

Eden, J.G., Greene, J.E., Osmundsen, J.F., Lubben, D., Abele, C.C.,
Gorbatkin, S. and Desai, H.D.

SEMICONDUCTOR THIN FILMS GROWN BY LASER PHOTOLYSIS

Thin ($\leqslant 1.2 \mu m$) Ge and Si films have been grown with rates up to
$3.6 \mu m/hr$ by laser-induced chemical vapor deposition (LCVD) on a
variety of substrates. Germanium films grown on amorphous SiO_2
(quartz) by photodissociating GeH_4 in He at 248nm (KrF laser)
exhibit grain sizes of 0.3-0.5 μm that increase only slightly up to
the pyrolytic threshold for GeH_4 (280°C). On (100) NaCℓ, however,
Ge films grown at a substrate temperature of 120°C are epitaxial.
The activation energy for LCVD growth of Ge films (from GeH_4) on
SiO_2 is measured to be 85 ± 20 meV which suggests that germanium is
arriving at the substrate in atomic form. The wavelength and
intensity dependence of the initial film growth rate supports the
conclusion that this process is photolytic and is initiated by the
absorption of a single photon.

Reprinted by permission of the publisher from "Laser Diagnostics and
Photochemical Processes for Semiconductor Devices", Eds. R.M.Osgood,
S.R.J.Brueck and H.R. Schlossberg, Mat.Res.Soc.Symp.Proc. 17:185,

Ehrlich, D.J. and Tsao, J.Y.

SPATIAL-RESOLUTION LIMITS OF LASER PATTERNING: SUBMICROMETER
PROJECTION MICROCHEMISTRY

The predictions of a simple two-dimensional model for the
sensitivity to process contrast of direct-write and projection laser
microchemistry are reviewed. Microchemical reactions excited by UV
image projection are demonstrated using a 30X reduction reflecting
system and excimer laser illumination. Well-resolved images with
submicrometer features are obtained by: (1) reaction of an Al/O
cermet, (2) projection doping of Si in BCl_3, (3) projection etching
of Pyrex glass in H_2, and (4) reactive patterning of an organic
bilayer. Linewidths of $0.5-0.2 \mu m$ have been demonstrated for
projection imaging in these systems.

Reprinted by permission of the publisher from "Laser-controlled
Chemical Processing of Surfaces", Eds. A.W.Johnson, D.J.Ehrlich,
H.R.Schlossberg, Mat.Res.Symp.Proc. 29:195, 1975. Copyright
Elsevier Science Publishing Co.Inc.

van Erk, W., van Hoek-Martensand, H.J.G.J. and Bartels, G.

THE EFFECT OF SUBSTRATE ORIENTATION ON THE GROWTH KINETICS OF GARNET
LIQUID PHASE EPITAXY

The growth kinetics of liquid-phase epitaxy of $Y_3Fe_5O_{12}$
substrates having (110), (211), (321), (100) or (111) orientation
has been studied and the results are discussed in terms of simple
theoretical growth models. The (111) face is a rough face on an
atomic scale, but the other faces are more or less singular faces.
Growth hillocks are observed on the surfaces of films grown on
(111), (211) and (321) substrates. The epitaxial layer in between
the hillocks grows via the nucleus-above-nucleus mechanism. Due to
the occurrence of two-dimensional nucleation on the steps of the
growth spiral, the slope of the hillocks is not a simple function of
the supersaturation. The mean distance for surface diffusion is
estimated to be twenty times the step height. Based on the growth
rate data the order of decreasing α-factor is (110), (211), (321),
(100), ... The results for growth on (111) substrates have been used
to estimate the effect of volume diffusion on the supersaturation.

Reprinted with permission from J.Cryst.Growth, 1980, 48:621,
published by North Holland Publishing Co., Amsterdam.

Eversteyn, F.C., Severin, P.J.W., v.d.Brekel, C.H.J. and Peek, H.L.

A STAGNANT LAYER MODEL FOR THE EPITAXIAL GROWTH OF SILICON FROM
SILANE IN A HORIZONTAL REACTOR

Flow patterns have been made visible in a horizontal water-cooled
epitaxial reactor by injection of TiO_2 particles into the gas flow.
From these experiments, a stagnant layer model has been developed
with which the epitaxial growth of silicon from silane can be
described. In the case of a nontilted susceptor, the model predicts
an appreciable nonuniformity in thickness along the susceptor,
whereas a small angle of tilting of the susceptor should yield a
much better uniformity in thickness (2% over a length of 22 cm).
Experiments agree very well with the theoretical predictions of the
model.

J.Electrochem.Soc. (1970), 117:125. Reprinted by permission of the
publisher, The Electrochemical Society Inc., Pennington, NJ, USA.

Farrow, R.F.C., Sullivan, P.W., Williams, G.M., Jones, G.R. and
Cameron, D.C.

MBE-GROWN FLUORIDE FILMS: A NEW CLASS OF EPITAXIAL DIELECTRICS

The deposition of cubic fluorite structure group II fluorides
onto several well known semiconductors has been studied under clean,
controlled, monitored conditions in an MBE system. The fluoride
beams were generated from a single Knudsen effusion oven for each
fluoride. Modulated beam mass spectrometry (MBMS) confirmed that
the fluorides sublimed as group II molecules. Fluoride films
therefore grow by a simple process of molecular condensation from
the fluoride beam. BaF_2 films deposited onto semiconductor surfaces
held at room temperature were polycrystalline. However, growth at
temperatures $\geqslant 200°C$ onto ion-bombarded, annealed InP and CdTe
surfaces resulted in twin-free epitaxial films with the parallel
epitaxial relation:

$$(001)BaF_2 \| (001)InP, \; CdTe,$$
$$[100]BaF_2 \| [100]InP, \; CdTe$$

Growth onto InP(001) surfaces subjected to no ion-bombardment
annealing treatment also resulted in twin free epitaxial films with
the same epitaxial relation. *In situ* electron diffraction studies
during fluoride film growth combined with *ex situ* studies suggest
that film growth occurs by a 2D mechanism rather than a 3D island
mechanism. Electrical measurements of BaF_2 and CaF_2 films show
typical film resistivities of $\geqslant 10^{13}$ Ω cm at 77 and 300K. Typical
breakdown fields are $\sim 5 \times 10^5 V$ cm^{-1} and promising interface properties
for MIS(BaF_2 insulator) structures on p-type Si and n-type

$Hg_{0.79}Cd_{0.21}Te$ have been achieved. Cubic fluorite structure group
II fluorides form a promising new class of epitaxial dielectrics
which can be grown at low (~200°C) substrate temperatures.

Reprinted with permission from J.Vac.Sci.Technol., 1981, B19:415,
published by the American Institute of Physics, New York © 1981,
American Vacuum Society.

Farrow, R.F.C., Wood, S., Greggi Jnr., J.C., Takei, W.J., Shirland,
F.A., and Furneaux, J.

A STUDY OF THE GROWTH CONDITIONS NECESSARY FOR REPRODUCIBLE
PREPARATION OF HIGH PERFECTION CdTe FILMS ON InSb BY MBE

The critical role of growth-related parameters including
substrate preparation conditions, source conditioning, growth rate
and growth temperature has been identified in MBE growth of
high-perfection CdTe films on InSb from a single CdTe effusion
source.

Reprinted with permission from J.Vac.Sci.Technol., 1985, B3(2):681,
published by the American Institute of Physics, New York. © 1985,
American Vacuum Society.

Fathauer, R.W., Lewis, N., Schowalter, L.J. and Hall, E.L.

ELECTRON MICROSCOPY OF EPITAXIAL $Si/CaF_2/Si$ STRUCTURES

Epitaxial $Si/CaF_2/Si(111)$ structures have been grown by
molecular beam epitaxy. Planar and cross-sectional transmission
electron microscopy has been used for characterisation of these
structures. The CaF_2 layer is high quality, single crystal material
exhibiting type B epitaxy with respect to the Si substrate. On the
other hand, the Si overgrowth layer consists of a mixture of
crystallites which are aligned either with the Si substrate or with
the CaF_2. The defect density in the overgrown Si improves with
distance from the interface with the CaF_2. Variation of the
substrate temperature during growth has relatively little effect on
the quality of the epitaxial Si.

Reprinted with permission from J.Vac.Sci.Technol. 1985, B3(2):736,
published by the American Institute of Physics, © 1985, American
Vacuum Society.

Faurie, J.P., Million, A., and Piaguet, J.

CdTe-HgTe MULTILAYERS GROWN BY MOLECULAR BEAM EPITAXY

Monocrystalline multilayers CdTe-HgTe have been grown for the first time using the molecular beam epitaxy (MBE) technique. A multilayer consisting alternately of CdTe (44A thick) and HgTe (180A thick), repeated 100 times, has been grown at 200°C with a good crystal quality. As for epilayers grown by MBE the crystallinity of the multilayers is improved by a raise in substrate temperature; moreover, the crystal quality is higher for HgTe than for CdTe up to 200°C. An upper limit of 40A for the interdiffusion depth between HgTe and CdTe layers has been determined from Auger electron spectroscopy and ion microprobe profiling measurements carried out on a $13\frac{1}{2}$-period multilayer, each period consisting of 150A for the CdTe layer and 400A for HgTe layer. We have also observed that no decrease occurs in the peak-to-valley ratio Auger signals if the focus of the electron gun is maintained at the center of the ion crater, during all the analysis, using the secondary electron image.

Reprinted with permission from Appl.Phys.Lett., 1982, 41(8):713, published by the American Institute of Physics, New York.

Feldman, R.D., Kisker, D.W., Austin, R.F., Jeffers, K.S. and Bridenbaugh, P.M.

A COMPARISON OF CdTe GROWN ON GaAs BY MOLECULAR BEAM AND ORGANOMETALLIC VAPOR PHASE EPITAXY

CdTe has been grown on (100)GaAs by molecular beam epitaxy (MBE), organometallic vapor phase epitaxy (OMVPE), and photo-assisted OMVPE (POMVPE). Optimum conditions for OMVPE are governed by the thermodynamics of the Cd-Te system. The conditions for growing (100) or (111) material for each technique are presented. (111) growth is shown to proceed via the formation of a relatively Te-poor Ga-As-Te surface phase. Surface treatments which result in different surface structures and yield (100) growth are presented.

Reprinted with permission from J.Vac.Sci.Technol., 1986, A4(4):2234, published by the American Institute of Physics, New York, © 1986 American Vacuum Society.

Fink, H.W., and Ehrlich, G.

DIRECT OBSERVATION OF OVERLAYER STRUCTURES ON W(110)

Observations in the field ion microscope have resolved the individual atoms in two-dimensional overlayers of platinum, palladium, iridium and silicon on W(110). All these layers, but most notably silicon, have quite an open arrangement of atoms. With iridium, transformations in the structure of the layer occur as the number of adatoms is changed.

Reprinted with permission from Surf.Sci., 1981, 110:L611, published by North Holland Publishing Co., Amsterdam.

Fischer R., Henderson, T., Klem, J., Masselink, W.T., Kopp, W., Morkoc, H.

CHARACTERISTICS OF GaAs/AlGaAs MODFETS GROWN DIRECTLY ON (100) SILICON

     We have successfully grown GaAs/AlGaAs MODFETs directly on silicon substrates by molecular beam epitaxy. We have found that an orientation slightly off (100) is well suited for the growth of GaAs/AlGaAs on silicon substrates. MODFETs fabricated from layers grown on Si had transconductances of 170 mS/mm at room temperature and exhibited no looping. When cooled to 77 K, the trans-conductance rose to 275 mS/mm. Hall mobilities of 51000 and 38000 cm^2/V s were obtained at 10 and 77 K respectively, with sheet electron densities of 8.30×10^{11} cm^{-2} in both cases. These results clearly demonstrate that device quality GaAs/AlGaAs is obtainable directly on Si substrates which has great implications with regard to the monolithic integration of III-V and Si technology.

Reprinted with permission from Electronics Lett., 1984, 20(22):945, published by the Institution of Electronics Engineers, Stevenage, UK.

Fujii, E., Nakamura, H., Haruna, K. and Koga, Y.

A QUANTITATIVE CALCULATION OF THE GROWTH RATE OF EPITAXIAL SILICON FROM SiCl$_4$ IN A BARREL REACTOR

     A model is proposed for the epitaxial growth of silicon in a vertical reactor where the reactant gas flows parallel to silicon wafers. It is assumed that the growth rate of deposited silicon is mass-transport controlled, and the equilibrium reaction is SiCl$_4$ + 2H$_2$ = Si + 4HCl. The model considers the profiles of the gas velocity and the temperature of the reactant gas, and also a decrease of the concentration along the gas flow. The growth rates at various positions along the susceptor are calculated under the conditions of substrate temperatures 1100°-1300°C, wall temperatures 300-850°C, reactant gas flow rates 40-120 liters/min, and concentrations of SiCl$_4$ to H$_2$ 0.004-0.008 in mole ratio. The growth rate is expressed as a function of the dimensions of the reaction chamber, and susceptor and of process variables such as the concentration, the gas flow rate and the temperature. Using the function, the sensitivity analysis of process variables to the

growth rate is investigated. Good agreement between the theory and
the experimental result of the growth rate is obtained.

J.Electrochem.Soc. (1972), 119:1106. Reprinted by permission of the
publisher, The Electrochemical Society Inc., Pennington, NJ, USA.

Giess, J., Gough, J.S., Irvine, S.J.C., Blackmore, G.W., Mullin,
J.B. and Royle, A.

THE GROWTH OF HIGH QUALITY $Cd_xHg_{1-x}Te$ BY MOVPE ONTO GaAs SUBSTRATES

The growth of $Cd_xHg_{1-x}Te$ layers by metalorganic vapour phase
epitaxy (MOVPE) onto (100) 2° → (110) GaAs substrates is reported.
The mirror smooth epitaxial layers have been grown with a
reproducible structural quality that is comparable to layers grown
onto the best CdTe substrates. By growing a sufficiently thick
buffer layer to isolate the active layer from the substrate, the
potential problem of Ga diffusion out of the substrate into the
layer has been successfully controlled. As a consequence the Ga
concentration in the active layer has been reduced to a level well
below that for normal background contamination.

Reprinted with permission from J.Cryst.Growth, 1985, 72:120,
published by North Holland Publishing Co., Amsterdam.

Giling, L.J.

GAS FLOW PATTERNS IN HORIZONTAL EPITAXIAL REACTOR CELLS OBSERVED BY
INTERFERENCE HOLOGRAPHY

Interference holography is used to visualise gas flow patterns
and temperature profiles in epitaxial systems. It is demonstrated
that in water-cooled horizontal reactor cells the carrier gases H_2
and He give dynamically stable laminar flow profiles throughout the
reactor. There is no indication of a stagnant or boundary layer for
flow velocities up to 80cm/sec in this type of cell. In air-cooled
cells, H_2 and He also give stable laminar flow profiles, but beyond
velocities of 40 cm/sec a cold gas finger appears in these flows due
to undeveloped flow and temperature profiles. In contrast to the
stable flow characteristics of H_2 and He, the flows of N_2 and Ar
always are unstable due to convective motions. Besides this
intrinsic instability, these flows are accompanied by severe
entrance effects (especially undeveloped flow profiles), which
dominate the flows for flow rates higher than 4 cm/sec. This is
observed for both air- and water-cooled cells. Another phenomenon
which is observed for N_2 and Ar is that beyond 4 cm/sec the
convective gas breaks up into a thin (8mm) laminar layer close to
the susceptor across which the entire temperature gradient is

present and a highly turbulent/convective part above this laminar layer. Analysis shows that for Ar and N_2 about eight times longer entrance lengths in the reactor cell are needed to achieve fully developed velocity and temperature profiles as compared with H_2 and He. This explains the dominant influence of this effect on the flows of Ar and N_2. When the influence of the entrance effect on the profiles is taken into account, all the observed flow patterns and temperature gradients are in agreement with the theoretical flows which can be predicted on basis of the respective Reynolds and Rayleigh (or Grashof) numbers.

J.Electrochem.Soc. (1982), 129(3):634. Reprinted by permission of the publisher, The Electrochemical Society Inc., Pennington, NJ USA.

Gilmer, G.H. and Broughton, J.Q.

SIMULATION MODELS OF THE CRYSTAL-VAPOR INTERFACE

The atomic structures of crystal-vapor surfaces have been studied using Monte Carlo simulations of the Ising model and molecular dynamics simulations of a system of Lennard-Jones particles. The roughening transition and its effects on the kinetics of growth and the crystal structure are discussed together with some implications for the growth of composition modulated crystals. Growth by two-dimensional nucleation and an impurity mechanism are discussed. Atomic positions at close-packed crystal-vapor surfaces are obtained from the molecular dynamics model. The atomic mobility in the surface region and the possibility of a surface melting transition are discussed.

Reprinted with permission from J.Vac.Sci.Technol., 1983, B1(2):298, published by the American Institute of Physics, © 1983, American Vacuum Society.

Haigh, J.

MECHANISMS OF METALLO-ORGANIC VAPOR PHASE EPITAXY AND ROUTES TO A ULTRAVIOLET-ASSISTED PROCESS

Ultraviolet irradiation during deposition has been reported to improve the morphology and increase the growth rate of GaAs epitaxial layers produced by metallo-organic vapor phase epitaxy. In order to optimise and control these effects for technological applications it is necessary to understand the mechanisms of metallo-organic vapor phase epitaxy (MOVPE). This paper reviews the current state of knowledge relevant to the UV intervention in InP growth. Preliminary experiments on the effect of UV in this system suggest that it may be increasing the mobility of the adsorbed reactant species.

Hardeman, R.W., Robbins, D.J., Gasson, D.B. and Daw, A.

OXIDE REMOVAL FROM SILICON WAFERS BY TRANSIENT MASS SPECTROMETRY AND X-RAY PHOTOELECTRON SPECTROSCOPY

We have studied the cleaning of silicon substrates in UHV prior to Si-MBE growth. Direct measurements have been made of the SiO species produced during thermal annealing and exposure to a silicon flux of substrates bearing a native oxide. Using these and XPS studies as a basis, we propose a model of a two stage process. In the first stage the oxide is reactively etched by the beam, while the second state is a bulk-like reaction (thermally activated E $\cong$ 3.9eV in the temperature range 770 to 850°C), occurring at the edges of oxide islands created on the surface. The second stage is the critical one for the clean-up process and can explain various observations of defect generation in epilayers grown at less than 800°C. We have also investigated the thermal desorption of the oxide produced by an RCA clean, which proceeds in the absence of a silicon flux over the same temperature range.

Hartman, P. and Perdok, W.G.

ON THE RELATIONS BETWEEN STRUCTURE AND MORPHOLOGY OF CRYSTALS. I.

An attempt is made to find relations between crystal structure and crystal morphology on an energy basis. It is concluded that the morphology of a crystal is governed by chains of strong bonds running through the structure. The effective period of such a chain of strong bonds is called a periodic bond chain vector (P.B.C. vector). The faces of a crystal are divided into three classes: (a) flat faces or F-faces, each of which is parallel to at least two P.B.C. vectors; (b) stepped or S-faces, each of which is parallel to at least one P.B.C. vector; (c) kinked faces or K-faces which are not parallel to any P.B.C. vector. F-faces are the most important faces; S-faces are of medium importance and K-faces are very rare or do not occur at all. Two examples are given, concerning the morphology of diamond and urotropine respectively.

Herman, M.A., Jylha, O. and Pessa, M.

ATOMIC LAYER EPITAXY OF $Cd_{1-x}Mn_xTe$ GROWN ON CdTe (III)B SUBSTRATES

Atomic Layer Epitaxy (ALE) has been applied for the first time to grow ternary compound semiconductor films, the constituent elements of which do not evaporate congruently. These films are $Cd_{1-x}Mn_xTe$ wide-gap semi-magnetic semiconductors with $x \leqslant 0.3$ deposited onto single crystal CdTe (111)B substrates.

Reprinted with permission from J.Cryst.Growth, 1984, <u>66</u>:480, published by North Holland Publishing Co., Amsterdam.

Hersee, S.D. and Duchemin, J.P.

LOW-PRESSURE CHEMICAL VAPOR DEPOSITION

In this chapter we review some of the major practical situations in which low pressure is used during the growth of semiconducting materials by chemical vapor deposition. Growth at subatmospheric pressure offers significant advantages for silicon grown from silane, and for the III-V compounds InP, GaAs, GaInAs, and GaInAsP grown from organometallic sources by vapor phase epitaxy (OM-VPE).

To understand the effect of pressure on growth it is necessary to have at least a qualitative knowledge of the physical and chemical processes involved in chemical vapor deposition (CVD). We therefore present a model of growth and of impurity incorporation that was initially developed for silicon growth but has since been applied, with some success, to the growth of compound semiconductors by OM–VPE. For each semiconductor we consider the practical growth system and then describe the material characteristics and applications in state-of-the-art devices.

Reproduced with permission from the Annual Review of Materials Science, Volume 12 © 1982 by Annual Reviews Inc.

Hersee, S.D., Baldy, M. and Duchemin, J.P.

THE VARIATION OF THE P/N JUNCTION POSITION IN GaAs/GaAlAs DOUBLE HETEROSTRUCTURES GROWN BY LOW PRESSURE MO VPE

A modified contact resistance profiler has been used to determine the doping type of the active layer in GaAs/GaAlAs double heterostructures grown by low pressure metal-organic vapour phase epitaxy (LP-MOVPE). It was found that under normal growth conditions, the p/n junction is located inside the "p-type"

$Ga_{0.65}Al_{0.35}As$ confinement layer. SIMS data show that this
displacement of the p/n junction is due to electrical compensation
of the zinc doping.

Reprinted with permission from J.Electron.Mater., 1983, 12(2):345,
published by The Metallurgical Society of AIME, Warrendale, PA, USA.

Hitchman, M.L.

A CONSIDERATION OF THE EFFECT OF THE THERMAL BOUNDARY LAYER ON CVD
GROWTH RATES

In deriving theoretical transport controlled growth rates for CVD
processes the effect of the temperature gradient close to the
reaction surface is usually neglected in order to obtain a simple
analytical treatment, and values of gas properties corresponding to
some average temperature are taken. A simple consideration of the
differential equations for fluid flow and heat and mass transfer
shows that one might expect the steep temperature profile commonly
encountered at a susceptor surface to lead to a considerable
distortion of the diffusion boundary layer and hence of the growth
rate. Nevertheless, the agreement between theory based on constant
gas property conditions and experiment for various reactor
geometries is usually reasonable. In this paper we present an
analysis to show why this is so.

Reprinted with permission from J.Cryst.Growth, 1980, 48:394,
published by North Holland Publishing Co., Amsterdam.

Hitchman, M.L. and Curtis, B.J.

HETEROGENEOUS KINETICS AND MASS TRANSFER IN CHEMICAL VAPOUR
DEPOSITION PROCESSES: III THE ROTATING DISC REACTOR

    A rotating disc reactor for the study of mass transport in CVD
processes is described. A simple theory for the transport is given
and the failure of experimental results obtained for the epitaxial
deposition of Si from $SiCl_4$ to conform to this theory is discussed.
By monitoring temperature profiles close to the disc surface, it is
shown that this failure arises from the fact that the fluid dynamics
in the reactor are determined not just by the disc rotating but by
the inlet gas flow as well. An empirical solution to the problem is
to adjust the experimental conditions so that the temperature
profiles at given rotation speeds agree with those expected on the
basis of the rotating disc theory. Growth rates of CVD layers under
these conditions are then found to be also in accord with theory.
With a knowledge of the transport contribution to growth, the

kinetic factors influencing the growth can then be examined.

Reprinted with permission from J.Cryst.Growth, 1982, 60:43, published by North Holland Publishing Co., Amsterdam.

Holmes, D.E., Chen, R.T., Elliott, K.R. and Kirkpatrick, C.G.

SYMMETRICAL CONTOURS OF DEEP LEVEL EL2 IN LIQUID ENCAPSULATED CZOCHRALSKI GaAs

We have determined the isoconcentration contours of the deep level EL2 across 3-in.diam, semi-insulating GaAs crystals grown by the liquid encapsulated Czochralski technique. The contours are essentially fourfold symmetric at the seed end of the crystals. The symmetrical pattern is independent of the melt stoichiometry and the relative direction of crystal and crucible rotation. EL2 distributions in the tail of the same crystals are often of lower symmetry. The results support a native defect model for EL2 in which the formation of the defect is enhanced by stress in the crystal.

Reprinted with permission from Appl.Phys.Lett., 1983, 43(3):305, published by the American Institute of Physics, New York.

Hottier, F. and Theeten, J.B.

SURFACE ANALYSIS DURING VAPOUR PHASE GROWTH

Several systems are described which enable surface analysis of a sample to be made during its processing in a vapour phase epitaxy chamber, GaAs homoepitaxy, using the H_2-GaCl-As$_4$ species, is discussed as a surface limited system and evidence of an adsorbed Ga-As-Cl complex is given for GaAs (111) and (111) surfaces. The low pressure chemical vapour deposition of Si on Si_3N_4 is shown to correspond to a nucleation-coalescence growth sequence. A detailed study of the formation and the time evolution of small Si nuclei (less than 100A in diameter) is given. GaAlAs/GaAs heterostructures, grown with organometallics, are analysed during their formation and interface anomalies are monitored and discussed.

Reprinted with permission from J.Cryst.Growth, 1980, 48:644, published by North Holland Publishing Co., Amsterdam.

Houzay, F., Moison, J.M. and Bensoussan, M.

ALUMINUM GROWTH ON (100) INDIUM PHOSPHIDE

The first detailed study of the room temperature growth of Al
on the 4x2(100)InP surface is reported. In the monolayer range, Al
strongly reacts with the substrate and an Al-In exchange reaction
occurs, leaving AlP clusters and metallic In. Above a coverage of
several monolayers, 3D single-crystal islands of Al grow. Their
epitaxial relationship with the substrate is (110)Al(100)InP, with
two possible domains. After a few hundred monolayers, the majority
domains coalesce, leading to a smooth (110)Al surface. The final
structure is formed by a thin AlP interface between bulk InP and the
Al film, whose surface is covered with seregated In.

Reprinted with permission from J.Vac.Sci.Technol., 1985, B3(2):756,
published by the American Institute of Physics, © 1985, American
Vacuum Society.

Hurle, D.T.J.

REVISED CALCULATION OF POINT DEFECT EQUILIBRIA AND NON-STOICHIOMETRY
IN GALLIUM ARSENIDE

From an analysis of relevant published literature, it is
deduced that the dominant disorder in GaAs is Frenkel defect
formation on the arsenic sub-lattice with the arsenic vacancy
exhibiting donor-like behaviour at high temperatures. Standard
methods of chemical thermodynamics are used to derive expressions
for the concentrations of the neutral and charged point defects.
Mass action constants are obtained from a combination of *a priori*
estimates and fitting to experimental data. The phase extent of
GaAs is mapped and the model shown to account adequately for the
density, lattice parameter and internal friction variations in
as-grown and heat treated material as well as for the high
temperature Hall coefficient behaviour.

J.Phys.Chem.Solids (1979), 40:613. Reprinted with permission from
the publishers, Pergamon Press, Oxford, UK.

Hurle, D.T.J.

SOLUBILITY AND POINT DEFECT-DOPANT INTERACTIONS IN GaAs - I.
TELLURIUM DOPING

The point defect model described in the companion paper is used
to construct solubility relations, predict the effects of heat
treatment and account for the large lattice dilations in tellurium
doped GaAs.

The complexing of Te donors with gallium vacancies is shown to
be very important and can be used to explain electrical

compensation, the formation of interstitial dislocation loops and the "superdilation" of n^+ material.

J.Phys.Chem.Solids, 1979, 40:627. Reprinted with permission from the publishers, Pergamon Press, Oxford, UK.

Hurle, D.T.J.

SOLUBILITY AND POINT DEFECT-DOPANT INTERACTIONS IN GaAs - II. TIN-DOPING

The form of the solubility curves for tin in both LPE and VPE grown GaAs are rationalised by a thermodynamic model which presumes that the dominant acceptor state due to tin is the donor-gallium vacancy complex $Sn_{Ga}V^-_{Ga}$ rather than the commonly postulated Sn^-_{As}.

In LPE growth from tin-rich solutions where the arsenic concentration in the melt exceeds the gallium concentration, very low electron mobilities have been reported and this is shown to arise from auto-compensation of the Sn^+_{Ga} donors by the $Sn_{Ga}V^-_{Ga}$ acceptors.

J.Phys.Chem.Solids, 1979, 40:639. Reprinted with permission from the publishers, Pergamon Press, Oxford, UK.

Hurle, D.T.J.

SOLUBILITY AND POINT DEFECT-DOPANT INTERACTIONS IN GALLIUM ARSENIDE - III

The form of the solubility curves for VPE and LPE grown material are rationalised using a thermodynamic model which predicts that compensation of donors in VPE material is due to the donor-gallium vacancy complex $Ge_{Ga}V^-_{Ga}$ whereas, in LPE material, the dominant acceptor making the material p type is the arsenic substituted state Ge^-_{As}. Mass action constants for the incorporation reactions for these entities and for the donor Ge^+_{Ga} are obtained. The compensation ratio in crystals grown from a congruent melt is predicted to be $N_D/N_A \sim 5.9$ and the transition to p type conductivity as one grows from Ga-rich solutions at progressively lower temperature is shown to occur at less than 100°C below the congruent melting point so that all LPE material should be p type. The superlinear behaviour of the total-germanium solubility curve at very high doping levels is postulated to be due to the formation of neutral $Ge_{Ga}Ge_{As}$ pairs.

J.Phys.Chem.Solids, 1979, 40:647. Reprinted with permission from the publishers, Pergamon Press, Oxford, UK.

Hurle, D.T.J.

THE SITE DISTRIBUTION OF AMPHOTERIC DOPANTS IN MULTIPLY-DOPED GaAs

A chemical thermodynamic approach is used to predict the site distribution and solubility of amphoteric dopants in multiply-doped crystals of GaAs. The analysis is applied to the experimental results of Greene and of Brozel et al on GaAs crystals doped with Si and either Sn, Ge or Se. Excellent agreement with the electrically measured carrier concentrations is obtained. It is shown that, whilst Ge commonly behaves as a donor in singly doped GaAs, in the presence of a high silicon concentration it can contribute more acceptors than donors to the crystal.

Reprinted with permission from J.Cryst.Growth, 1980, 50:638, published by North Holland Publishing Co., Amsterdam.

Hurle, D.T.J.

CONVECTIVE TRANSPORT IN MELT GROWTH SYSTEMS

Most melt growth systems are thermal transport controlled, ie the kinetic undercooling of the growth surface and the depression of the freezing point due to the added solute are both small. In consequence, the microscopic growth rate and the solute incorporation are very sensitive to time variation in the melt temperature. In this review several aspects of this problem are considered but first the basic pattern of flow in the Czochralski melt is discussed. The spoke pattern which occurs on oxide melts is described and possible explanations reviewed. The catastrophic flow transitions observed in two particular oxide melts are described. Some general considerations of time dependent flow are given and a detailed description of such flow in a floating zone is provided. Finally, recent theoretical and experimental studies of the coupling between morphological and convective instabilities are described.

Reprinted with permission from J.Cryst.Growth, 1983, 65:124, published by North Holland Publishing Co., Amsterdam.

Hwang, G.J. and Cheng, K.C.

CONVECTIVE INSTABILITY IN THE THERMAL ENTRANCE REGION OF A HORIZONTAL PARALLEL-PLATE CHANNEL HEATED FROM BELOW

An investigation is carried out to determine the conditions marking the onset of longitudinal vortex rolls due to buoyant forces in the thermal entrance region of a horizontal parallel-plate

channel where the lower plate is heated isothermally and the upper
plate is cooled isothermally. Axial heat conduction is included in
an analytical solution for the Graetz problem with fully developed
laminar velocity profile. Linear-stability theory based on
Boussinesq approximation is employed in the derivation of
perturbation equations. An iterative procedure using high-order
finite-difference approximation is applied to solve the perturbation
equations and a comparison is made against the conventional
second-order approximation. It is found that for $Pr \geqslant 0.7$ the flow
is more stable in the thermal entrance region than in the fully
developed region, but the situation is just opposite for small
Prandtl number, say $Pr \leqslant 0.2$. Graphical results for the critical
Rayleigh numbers and the corresponding disturbance wavenumbers are
presented for the case of $Pe \rightarrow \infty$ with Prandtl number as a parameter
and the case of air ($Pr = 0.7$) with Peclet number as a parameter in
the range of dimensionless axial distance from the entrance between
$x = 0.001$ and 4×10^{-1}.

Reprinted with permission from J.Heat Transfer, 1973, 95:72,
published by the American Society of Mechanical Engineers, New York.

Ihara, M., Arimoto, Y., Jifuku, M., Kimura, T., Kodama, S.,
Yamawaki, H. and Yamaoka, T.

VAPOR PHASE EPITAXIAL GROWTH OF MgO/Al_2O_3

An Si on $MgO.Al_2O_3$ on Si double heterostructure material for
integrated circuits has been developed. The $MgO.Al_2O_3$ epitaxial
layer on Si was grown by using an open-tube $Al-HCl-MgCl_2-CO_2-H_2$
vapor phase transport technique. The electron Hall mobility and the
defect density of the Si active layer on epitaxially grown $MgO.Al_2O_3$
were $\mu_n = 1400cm^2/Vsec$ and $1 \times 10^4/cm^2$, respectively, when the Si
thickness was $8\mu m$ and the doping concentration was $n=2x10^{14}/cm^3$.
The growth conditions of $MgO.Al_2O_3$ epitaxial layers and the
characteristics of the Si active layers are presented.

J.Electrochem.Soc. 1982, 129:2569. Reprinted by permission of the
publishers, The Electrochemical Society Inc., Pennington, NJ, USA.

Irvine, S.J.C., Mullin, J.B. and Tunnicliffe, J.

PHOTOSENSITISATION: A STIMULANT FOR THE LOW TEMPERATURE GROWTH OF
EPITAXIAL HgTe

Epitaxial growth of HgTe layers on CdTe and InSb substrates is
reported where ultraviolet (UV) radiation is used to dissociate
diethyltelluride (Et_2Te). Good quality single crystal growth has
been achieved over a range of temperatures from 200 to 300°C, well

below temperatures expected to pyrolytically dissociate Et$_2$Te. It was found that high Hg pressures (>1x10^{-2}atm) were needed to successfully dissociate Et$_2$Te and grow HgTe. Mechanisms for the dissociation of Et$_2$Te are discussed and a model is proposed which relies on a photosensitisation mechanism. This model is based on a surface selective reaction which is necessary for good epitaxial growth and high surface quality. The crystalline quality of these epitaxial layers has been established using X-ray diffraction and electron channelling. Layer thickness and growth rates were determined using a Scanning Electron Microscope (SEM). The results of growth rate measurements on different substrate orientations are discussed together with growth rate dependence on temperature and Et$_2$Te concentration. It is found that surface kinetics as well as transport are important in this growth process.

Reprinted with permission from J.Cryst.Growth, 1984, 68:188, published by North Holland Publishing Co., Amsterdam.

Irvine, S.J.C., Giess, J., Mullin, J.B., Blackmore, G.W. and Dosser, O.D.

PHOTO-METAL ORGANIC VAPOR PHASE EPITAXY: A LOW TEMPERATURE METHOD FOR THE GROWTH OF Cd$_x$Hg$_{1-x}$Te

A new epitaxial growth technique, photo-metal organic vapor phase epitaxy (photo-MOVPE) is described for the growth of Cd$_x$Hg$_{1-x}$Te (CMT). Ultraviolet radiation is used to decompose the metal-organics, diethyl-telluride (Et$_2$Te) and dimethyl-cadmium (Me$_2$Cd) on the substrate surface. Epitaxial HgTe can be grown at relatively low temperatures (200-300°C) by a surface photosensitisation reaction. However, the growth of CMT is only epitaxial if an inert carrier gas (He) is used instead of H$_2$. These results are described in terms of suppression of vapour phase nucleation while allowing surface nucleation to occur. Vapour phase nucleation results in the deposition of fine particles of CMT and can totally disrupt epitaxial growth. The growth rate dependence of CMT on CdTe substrate orientation is considered and compared with HgTe growth rates on the same orientations. These results show that surface kinetics can dominate the growth process, making it insensitive to vapour concentrations. Secondary ion mass spectrometry (SIMS) profiles through a CMT layer, grown at 250°C, for the major elements, Cd, Hg and Te, show an abrupt interface between the substrate and layer of less than 400A. Carbon impurity profiles through HgTe layers show that using the inert gas, carbon incorporation is no greater than for H$_2$ carrier gas. It is shown that carbon is not a major contaminant in the epitaxial HgTe and CMT layers, but may still be a problem at lower concentrations.

Reprinted with permission from J.Vac.Sci.Technol., 1985, B3:1450,

Irvine, S.J.C., Mullin, J.B., Robbins, D.J. and Glasper, J.L.

A STUDY OF UV ABSORPTION SPECTRA AND PHOTOLYSIS OF SOME GROUP II AND GROUP VI ALKYLS

A preliminary study has been made of the UV photolysis of metal-organic compounds of Hg, Cd and Te which could be used for low temperature, selective area deposition of cadmium mercury telluride (CMT). High resolution UV absorption spectra have been measured for dimethylcadmium ($CdMe_2$), dimethylmercury ($HgMe_2$) and diethyltelluride ($TeEt_2$). Possible modes for photodissociation are discussed in light of these results. The photodissociation of these alkyls was attempted in a hydrogen stream at atmospheric pressure using a mercury-xenon lamp, deposition being onto a silica reaction tube. Yields of Cd, Hg and Te were measured under different deposition conditions to determine the dependence on UV intensity, alkyl concentration and flow velocity.

J.Electrochem.Soc. 1985, 132(4):968. Reprinted by permission of the publishers, The Electrochemical Society Inc., Pennington, NJ, USA.

Irvine, S.J.C., Giess, J., Gough, J.S., Blackmore, G.W., Royle, A., Mullin, J.B., Chew, N.G. and Cullis, A.G.

THE POTENTIAL FOR ABRUPT INTERFACES IN $Cd_xHg_{1-x}Te$ USING THERMAL AND PHOTO-MOVPE

This paper reviews the current progress in the growth of abrupt structures in the infrared detector alloy $Cd_xHg_{1-x}Te$, with special reference to metalorganic vapour phase epitaxy (MOVPE). Recent results on the growth of heterostructures using the interdiffused multilayer process (IMP) are described for epitaxy onto GaAs as well as CdTe substrates. It is envisaged that useful heterostructure devices can be grown where the interface widths are of the order of $0.3\mu m$. For more abrupt structures, lower growth temperatures are needed and this has been demonstrated using the new photolytic MOVPE process. Epitaxial growth at temperatures as low as 200°C has enabled measured interface widths of approximately 100 A to be realised for a HgTe/CdTe structure. Lower growth temperatures also reduce the rates of diffusion of dopants such as Ga from a GaAs substrate. Ga concentrations of just 0.05 ppma have been measured 500 A from a CdTe/GaAs interface. Detailed Hall measurements on photo-MOVPE HgTe and HgTe/CdTe structures have shown that high quality epitaxial layers can be grown. A study of the limitations on the electrical quality of using photo-MOVPE for the growth of

CdTe/HgTe superlattices has been explored by predicting the extent
of interdiffusion at 200°C and 150°C. Even at 150°C, the predicted
interdiffusion in just 10 min is significant. However,
interdiffusion may depend critically on the dislocation structure
and strain at the interface. Structural studies on thin epitaxial
layers shows the critical dependence of structure and strain on the
substrate orientation and layer thickness.

Reprinted with permission from J.Cryst.Growth, 1986, published by
North Holland Publishing Co., Amsterdam.

Ishiwara, H., Asano, T. and Furukawa, S.

EPITAXIAL GROWTH OF ELEMENTAL SEMICONDUCTOR FILMS ONTO SILICIDE/Si AND FLUORIDE/Si STRUCTURES

Epitaxial growth of Si and Ge films onto silicide/Si structures
has been reviewed. Growth conditions of epitaxial silicide films
such as Pd_2Si, $CoSi_2$ and $NiSi_2$ films on Si substrates, as well as
heteroepitaxial $Si/CoSi_2/Si$ structures, are presented. Crystalline
quality and structural properties of the films have been analysed by
Rutherford backscattering and channeling spectroscopy, transmission
electron microscopy, optical microscopy, and X-ray diffraction
analysis. Growth conditions of fluoride films such as CaF_2, SrF_2
and BaF_2 films on Si substrates are also presented. For the growth
of Si and Ge overlayers on the fluoride/Si structure, the lattice
matching condition between the semiconductor and fluoride films has
approximately been satisfied by use of pure and mixed fluoride films.

Reprinted with permission from J.Vac.Sci.Technol., 1983, B1(2):266,
published by the American Institute of Physics, New York © 1983,
American Vacuum Society.

Jackson, K.A.

MECHANISM OF GROWTH

A crystal grows when atoms are added to it. The way in which
atoms are added to the solid determines the morphology of the solid
and the number and distribution of imperfections in the solid.

A crystalline solid can grow from its liquid, from its vapor,
from a solution or from another solid. The character of the
crystallisation will depend on the phase into which the crystal is
growing and on other conditions such as the temperature, pressure,
concentration, etc. These conditions determine the atom movements
by which the growth of a crystal takes place.

The atom movements cannot be observed directly. In the present paper we will discuss two manifestations of the atom movements; namely, the rates of growth of crystals and the morphology of the crystal surfaces. These manifestations will be discussed in terms of atom movement, and related to the conditions of growth.

From 'Liquid Metals and Solidification', American Society for Metals, 1958, p174 - with permission.

Jackson, K.A.

THEORY OF MELT GROWTH

Crystal growth is a complex process which usually takes place by a phase change. This chapter will be concerned with the melt growth of crystals which takes place by a phase change from liquid to solid. An initial discussion on solid and liquid phases will be followed by a brief consideration of the equilibrium between solid and liquid. But, the bulk of this chapter will deal with the dynamics of the phase change.

Reprinted with permission from 'Theory of Melt Growth', p21, in 'Crystal Growth and Characterisation', Eds. R.Ueda and J.B.Mullin, North Holland, Elsevier Science Publishers, 1975.

James, T.W. and Stoller, R.E.

BLOCKING OF THREADING DISLOCATIONS BY $Hg_{1-x}Cd_xTe$ EPITAXIAL LAYERS

The process by which threading dislocations present in the substrate are bent into the interface plane during epitaxial layer growth and thus are blocked from propagating into the epitaxial layer has been observed in growth of the II-VI ternary alloy $Hg_{1-x}Cd_xTe$ on CdTe and $Cd_{1-z}Zn_zTe$ substrates. Stereo-pair transmission electron micrographs were analysed which clearly revealed the interfacial dislocations to be in a planar network, as has been previously reported for III-V epitaxial systems in which the blocking mechanism has been identified by other investigators. Additional evidence of the blocking phenomenon has been provided by chemical dislocation - revealing etching, which verified that the substrate dislocations were not present in the layer.

Reprinted with permission from Appl.Phys.Lett., 1984, $\underline{44}$(1):56, published by the American Institute of Physics, New York.

Jastrzebski, L., Corboy, J.F. and Pagliaro Jr., R.

GROWTH OF ELECTRONIC QUALITY SILICON OVER SiO_2 BY EPITAXIAL LATERAL
OVERGROWTH TECHNIQUE

The Epitaxial lateral Overgrowth (ELO) technique is described.
In this technique windows are cut in the SiO_2 mask overlying the
silicon substrate and a silicon film is deposited from a SiH_2Cl_2,
H_2, HCl mixture at 1000-1150°C. The silicon epilayer nucleates at
the oxide window and grows upwards and then spreads laterally across
the oxide surface to produce an Si-SiO_2-Si structure. One of the
problems of the technique is spurious nucleation of new growth
centres on the oxide surface. Ways of minimising this problem are
discussed.

Reprinted with permission from J.Cryst.Growth, 1983, 63:493,
published by North Holland Publishing Co., Amsterdam.

Jastrzebski, L.

SOI BY CVD: EPITAXIAL LATERAL OVERGROWTH (ELO) PROCESS - REVIEW

The ELO process (Epitaxial Lateral Overgrowth), based on the
growth of silicon film over an SiO_2 mask by the CVD technique, has
been reviewed. The silicon film is locally seeded in openings of
the SiO_2 mask from which it overgrows the oxide. The key to success
is selective deposition: growth conditions which give growth of
silicon over exposed silicon surfaces and at the same time prevent
nucleation of silicon over oxide. These conditions, together with
growth and surface morphology of obtained ELO films, are discussed
as a function of gas composition, growth temperature, growth
procedure and type of reactor. Defect structure and electric
properties of ELO films are also reviewed together with performance
of devices made in these films. Data obtained so far showed that
monocrystalline silicon films with smooth mirror-like surfaces and
low defect density can be grown over 20 μm wide SiO_2 islands by the
optimised ELO process. The performance of devices made in these ELO
films was similar to devices made in homoepitaxial layers. On the
basis of these data it is believed that the ELO technique is a
promising approach for the fabrication of SOI films.

Reprinted with permission from J.Cryst.Growth, 1983, 63:493,
published by North Holland Publishing Co., Amsterdam.

Jones, C.L, Capper, P. and Gosney, J.J.

THERMAL MODELLING OF BRIDGMAN CRYSTAL GROWTH

Electrical analogues have been used to model the thermal behaviour of a Bridgman crystal growing system. Computer analysis of the analogues yields detailed information on temperatures and heat flows in a complete system, ie furnace, water-jacket, ampoule and crystal. For the particular system modelled it is found that isotherm shapes within the crystal during the critical early stages of growth are strongly affected by the way heat is lost from the bottom of the ampoule. By changing the shape of the ampoule bottom or by changing the conductivity of the stem supporting the ampoule the isotherm shapes can be readily altered. The model also shows that the crystal growth rate is 15% slower than the ampoule lowering speed early in the growth cycle.

Reprinted with permission from J.Cryst.Growth, 1982, 56:581, published by North Holland Publishing Co., Amsterdam.

Jones, C.L., Capper, P., Gosney, J.J. and Kenworthy, I.

FACTORS AFFECTING ISOTHERM SHAPE DURING BRIDGMAN CRYSTAL GROWTH

Isotherm shape is an important factor controlling the quality of Bridgman grown crystals. A convex (relative to the solid) isotherm will often lead to the best crystallinity but if segregation is important a flat isotherm is desirable. A thermal model is used to establish the effects of temperature dependent conductivity, ampoule transparency and furnace temperature profile on heat flows and temperature distributions in the growing crystal. The shape of the first-to-freeze isotherm is found to depend on the furnace profile and the way heat is lost from the base of the melt. A convex isotherm requires a shallow profile and high end losses. As growth progresses end losses are less important and the growth isotherm shape is determined by the crystal-melt conductivity and the furnace profile.

Reprinted with permission from J.Cryst.Growth, 1984, 69:281, published by North Holland Publishing Co., Amsterdam.

Jordan, A.S.

AN EVALUATION OF THE THERMAL AND ELASTIC CONSTANTS AFFECTING GaAs CRYSTAL GROWTH

The thermal and elastic constants essential to a realistic modeling of GaAs crystal growth have been critically evaluated. High temperature values are recommended for the thermal expansion coefficient, elastic stiffnesses, thermal stress modulus, critical resolved shear stress, thermal conductivity and diffusivity of

GaAs. Furthermore, the radiative and convective heat transfer
coefficients, h, have been determined from Newton's Law of Cooling
in combination with the Stefan Boltzmann relation and a
hydro-dynamic approximation, respectively. The evaluation of these
coefficients has been facilitated by the recent calculation of the
total emittance of GaAs and the availability of the thermal and
transport properties of the ambient fluids. The coefficients h_{conv}
for B_2O_3 (1), He(g), N_2 (g), and A(g) and h_{rad} for a series of
doping levels are presented in a convenient graphical form. Due to
its low viscosity at elevated temperatures, h_{conv} for anhydrous
B_2O_3(1) is ~0.2cm^{-1} which is more than an order of magnitude larger
than that for any one of the gaseous ambients; h_{rad} increases with
ambient temperature and doping level, ranging from ~0.15 to 0.4cm^{-1}
at 1300K. This indicates that radiation and free convection via
B_2O_3(1) are competitive heat dissipating processes. The effect of
these parameters is illustrated by means of the quasi-steady state
heat transfer/thermal stress model for slip-induced dislocation
generation in GaAs. A novel plotting technique maps the thermal
stress corresponding to the {111}, {110} slip system in excess of
the critical resolved shear stress for a {100} wafer, a quantity
defined to be proportional to the dislocation density. It is shown
that the central area (~75%) of a 2cm diameter boule with 10^{16}cm^{-3}
n-type impurities grown in N_2 at an ambient temperature 20K below
the melting point would contain no dislocations. A lower ambient
temperature, B_2O_3 (1) encapsulation and/or an increase in doping
level leads to a progressive elimination of dislocation-free areas
with a concomitant rise in the dislocation density.

Reprinted with permission from J.Cryst.Growth, 1980, 49:631,
published by North Holland Publishing Co., Amsterdam.

Joyce, B.A., Foxon, C.T. and Neave, J.H.

FUNDAMENTALS OF MOECULAR BEAM EPITAXY

The paper deals with three fundamental aspects of epitaxy: the
surface chemistry of growth, the growth process and doping. In the
first of these, the adsorption mechanism for As_2 amd As_4 are
considered, together with the effects of As_2 loss at temperatures
above 600 K. Kinetic processes on the growth of GaAs are then
compared with those involved in the growth of $Ga_xIn_{1-x}As(P)$.

Single crystal growth of GaAs was obtained in the range 345-870
K. Between 870 and 530 K an atomically smooth reconstructed surface
is obtained. In the range 530 - 365 K the films show a bulk surface
structure, whilst below 365 K surface roughening occurs. Below 345
K the films were amorphous.

Finally, the doping characteristics of Be, Sn and Ge are

considered. Be behaves as an almost ideal p-type dopant. The incorporation of Sn is more complicated and results are presented to indicate a surface rate limitation which is strongly dependent on As_4 flux and, less strongly, on substrate temperature. Ge is an amphoteric dopant but incorporation on As sites to give acceptor behaviour is limited by the early formation of free Ga on the surface under Ga-rich growth conditions.

J.Japn.Assoc.Cryst.Growth, 1978, 5:185.

Joyce, B.A.

KINETIC AND SURFACE ASPECTS OF MBE

Surface kinetic data relating to the growth of thin films of III-V compounds and alloys by MBE can be obtained using modulated beam techniques. Data acquisition and signal processing methods are described, and application to the study of $Gas-As_4$ and $Ga-As_2$ surface reactions presented in detail. The evaluation of composition control during growth of both mixed Group III and mixed Group V element alloys is also discussed.

Some consequences on film properties of the differing surface chemistry involved in growth from Group V element tetramers compared with dimers are considered in relation to possible reaction mechanisms. The evaluation of the surface crystallographic and electronic structure of differently reconstructed GaAs(001) surfaces using RHEED, ARPES and core level spectroscopy is briefly reviewed.

Reprinted with permission from "Molecular Beam Epitaxy and Heterostructures", Eds. L.L.Chang and K.Ploog, NATO ASI Series E - Applied Sciences No.87, Martinus Nijhoff, Dordrecht, 1985.

Juza, J. and Cermak, J.

PHENOMENOLOGICAL MODEL OF THE CVD EPITAXIAL REACTOR

A model of the epitaxial chemical vapor deposition reactor has been developed based on fundamental physicochemical principles. The kinetics of the heterogeneous deposition reaction is taken into account along with transport phenomena occurring in the gas phase above the deposition surface. The obtained set of partial differential equations is solved numerically, the results being compared with a representative set of experimental data and with previously published models. The comparison demonstrates the suitability of the model conception as well as its applicability for process design purposes.

J.Electrochem.Soc. 1982, 129:1627. Reprinted by permission of the publishers, The Electrochemical Society Inc., Pennington, NJ, USA.

Kao, Y.C., Tejwani, M., Xie, Y.H., Lin, T.L. and Wang, K.L.

UNIFORMITY AND CRYSTALLINE QUALITY OF CoSi$_2$/Si HETEROSTRUCTURES GROWN BY MOLECULAR BEAM EPITAXY AND REACTIVE DEPOSITION EPITAXY

The crystal quality and the lateral uniformity of CoSi$_2$/Si epitaxial heterostructures grown by three different techniques have been investigated. The growth techniques used were molecular beam epitaxy (MBE), reactive deposition epitaxy and solid phase epitaxy (SPE). The growth of the silicide and of silicon epilayers was carried out under ultra high vacuum conditions. Crystal quality was determined by Rutherford backscattering spectroscopy (RBS), and the surface morphology was studied using scanning electron micrographs. The crystal quality and lateral uniformity of silicide films grown by MBE in a proper condition were better than those grown by the other two techniques. Silicide films with minimum RBS channeling yields of 3% were successfully obtained using MBE. Further effort is needed to improve the lateral uniformity of the silicide films for fabrication of Si/CoSi$_2$/Si epitaxial heterostructures.

Reprinted with permission from J.Vac.Sci.Technol., 1985, B3(2):596, published by the American Institute of Physics, © 1985, American Vacuum Society.

Kapitan, L.W., Litton, C.W., Clark, G.C. and Colter, P.C.

ON THE DESIGN AND CHARACTERISATION OF A NOVEL ARSINE CRACKING FURNACE UTILISING CATALYTIC DECOMPOSITION OF AsH$_3$ TO YIELD A PURELY MONOMERIC SOURCE OF ARSENIC FOR MOLECULAR BEAM EPITAXIAL GROWTH OF GaAs

Arsine (AsH$_3$) used as an arsenic source in molecular beam epitaxy instead of solid arsenic has been found to decompose catalytically instead of thermally under MBE growth conditions. Dehydrogenation of AsH$_3$ in a spiral quartz cracking furnace containing an electrically heated catalyst has produced a purely monomeric beam of arsenic suitable for the growth of MBE GaAs and related III-V compounds. Experimental procedures and results of the AsH$_3$ cracking are reported together with electrical and optical characterisation of the first purely As$_1$ grown MBE GaAs.

Reprinted with permission from J.Vac.Sci.Technol., 1984, B2(2):280 published by the American Institute of Physics, © 1984, American Vacuum Society.

Kaplan, R. and Bottka, N.

EPITAXIAL GROWTH OF Fe ON GaAs BY METALORGANIC CHEMICAL VAPOR
DEPOSITION IN ULTRAHIGH VACUUM

Fe epitaxial films have been grown on GaAs(100) by thermal
dissociation of $Fe(CO)_5$ in a high vacuum environment. *In situ*
low-energy electron diffraction (LEED) and Auger spectroscopy have
been used to study the MOCVD process and to characterise the growing
films. Excellent film quality is evidenced by the observed small
ferromagnetic resonance linewidth.

Reprinted with permission from Appl.Phys.Lett. 1982, 41(10):972
published by the American Institute of Physics.

Karlicek Jr., R.F., Donnelly, V.M. and Johnston Jr., W.D.

LASER SPECTROSCOPIC INVESTIGATION OF GAS-PHASE PROCESSES RELEVANT TO
SEMICONDUCTOR DEVICE FABRICATION

Chemical vapor deposition (CVD) and plasma etching are
important gas-phase techniques used in fabricating semiconductor
devices. These processes frequently involve poorly understood
multicomponent gas-phase reactions which control reproducibility and
product quality. Laser spectroscopic techniques have recently been
developed to investigate CVD and plasma etching. These methods
offer several advantages for probing complex systems. A comparison
of various probing techniques will be presented, and recent results
of laser spectroscopic investigations of plasma etching and CVD of
silicon and III-V compounds will be reviewed.

Reprinted with permission from "Laser Spectroscopic Investigation of
Gas-phase Processes Relevant to semiconductor Device Fabrication",
MRS Symp.Proc. 17:151. Copyright 1983 by Elsevier Science
Publishing Co.Inc., Amsterdam.

Kawaguchi, Y., Asahi, H. and Nagai, H.

GAS SOURCE MBE GROWTH OF HIGH-QUALITY InP USING TRIETHYLINDIUM AND
PHOSPHINE

High quality InP layers have been grown on (100) InP substrates
by gas source molecular beam epitaxy using triethylindium and
phosphine as III and V group sources. The electrical and optical
properties of grown InP films are evaluated and compared with those
of films grown using red phosphorus as a P source. Epitaxial layers
are n-type and the highest achieved 77K mobility is $24000cm^2/Vs$ with
a carrier concentration of $4.1 \times 10^{15} cm^{-3}$. When phosphine is used

instead of red phosphorus as a P source, the compensation ratio
(N_A/N_D) decreases from 0.60-0.98 to 0.29-0.31.

Reprinted with permission from Jpn.J.Appl.Phys. 1985, 24(4):L221

Kisker, D.W. and Feldman, R.D.

PHOTON ASSISTED OMVPE GROWTH OF CdTe

We have investigated the use of ultraviolet light for lowering
the temperature of Organometallic Vapor Phase Epitaxial (OMVPE) CdTe
growth on GaAs. The effects of growth temperature and stoichiometry
on the crystallinity are reported. In addition, we have observed
what appears to be anomalously high quantum efficiency for this
process, which suggests a more complex phenomena than simply direct
photolysis.

Reprinted with permission from J.Cryst.Growth, 1985, 72:102,
published by North Holland Publishing Co., Amsterdam.

Kobayashi, N.

HYDRODYNAMICS IN CZOCHRALSKI GROWTH - COMPUTER ANALYSIS AND
EXPERIMENTS

Probable flows in a melt are reviewed. In a pure melt, four
different flows, free convection, surface tension driven flow
(Marangoni flow), forced convection caused by crystal rotation, and
forced convection caused by crucible rotation, are to be
considered. The mixed flow caused both by buoyant force and by
crystal rotation is roughly divided into three classes with respect
to the ratio Gr/Re^2, where Gr is the Grashof number due to the
temperature difference between the crystal and crucible and Re is
the Reynolds number of the rotating crystal. For a low Gr/Re^2,
forced convection is dominant in the melt, while for a high Gr/Re^2,
free convection is dominant. For an intermediate Gr/Re^2, free and
forced convection coexist and they are separated by a stagnant
surface. The temperature field is slightly dependent on the flow
for low Prandtl number melts such as molten metals and
semiconductors, while it strongly depends on the flow for high
Prandtl number melts such as molten oxides. Two distinct types of
temperature field are obtained with respect to the dominant
convection modes in the melt: free convection and forced
convection. For a fixed Grashof number, the heat transfer mechanism
changes from free convection dominant mode to forced convection
dominant mode as the Reynolds number increases. This explains the
change of the interface shape with increasing crystal rotation rate
and the abrupt inversion of the interface shape with increasing

crystal diameter. A formula to predict the critical Reynolds number
at which the interface shape changes from convex to concave is
presented. The effect of heat dissipation from the melt free
surface on the flow is also examined, and a reason for the
difficulty encountered in the growth of some oxide crystals is
offered. Finally, the effect of the crucible rotation on the flow
is briefly described.

Reprinted with permission from J.Cryst.Growth, 1981, 52:425,
published by North Holland Publishing Co., Amsterdam.

Kolbesen, B.O.

CARBON IN SILICON

The properties and effects of carbon in silicon and its
behavior during device processing are reviewed. The typical carbon
concentrations in electronic-grade silicon are still on the order of
$10^{16}cm^{-3}$. Due to the small distribution coefficient, carbon is
incorporated inhomogeneously during crystal growth. This results in
considerable variations of the carbon concentration on a macroscopic
and microscopic scale (striations). Carbon strongly interacts with
intrinsic atomic point defects and impurities and plays a prominent
role as a nucleating center in the formation of microdefects in
float-zoned (FZ) and Czochralski (CZ)-grown silicon. In FZ silicon
carbon concentrations exceeding about $5x10^{16}cm^{-3}$ can cause the
formation of process-induced crystal defects in the production of
power devices. These defects are frequently arranged in a
swirl-like pattern and degrade the electrical characteristics of the
devices considerably. On the contrary high carbon concentrations in
FZ silicon behave rather benignly in the fabrication of integrated
circuits.

Reprinted with permission from "Aggregation Phenomena of Point
Defects in Silicon", p155. Eds. E.Sirtl and J.Goorisen. Published
by The Electrochemical Society Inc., Pennington, NJ, USA.

Koppitz, M., Vestvik, O. and Pletschen, W.

ON THE EFFECT OF CARRIER GAS ON GROWTH CONDITIONS IN MOCVD REACTORS;
RAMAN STUDY OF LOCAL TEMPERATURE

The temperature profiles in a MOCVD system have been studied by
means of rotational Raman scattering for different carrier gases and
flow rates. The temperature gradient normal to the substrate
surface was four times larger in N_2 than in H_2. It increases
rapidly with the flow rate through the reactor. Thus, at high N_2
concentrations or high gas velocities the width of the high

temperature region above the susceptor is drastically reduced. This
effect possibly explains the improved growth of InP at such
conditions.

Reprinted with permission from J.Cryst.Growth, 1984, 64:136,
published by North Holland Publishing Co., Amsterdam.

Krautle, H., Roentgen, P. and Beneking, H.

EPITAXIAL GROWTH OF Ge ON GaAs SUBSTRATES

   Epitaxial layers of Ge have been grown on GaAs(100) substrates
by thermal decomposition of GeH_4 in hydrogen at atmospheric
pressure. Growth was investigated for growth temperatures between
400 and 700°C at growth rates of 0.1-10μm/h. Two different growth
areas were observed. Above 500°C the growth is nearly constant and
mainly limited by gas phase diffusion. Below this temperature, the
growth rate decreases rapidly. For low temperature growth, an
activation energy of 1.4 eV was found. Layers deposited at high
temperature show good crystallinity but highly doped thin
interfacial layers due to Ga and As diffusion. This effect is
reduced at low temperatures. At 400°C smooth layers have been
deposited with a background doping below $n=10^{17}cm^{-3}$.

Reprinted with permission from J.Cryst.Growth, 1983, 65:439,
published by North Holland Publishing Co., Amsterdam.

Lagowski, J., Gatos, H.C., Parsey, J.M., Wada, K., Kaminska, M. and
Walukiewicz, W.

ORIGIN OF THE 0.82eV ELECTRON TRAP IN GaAs AND ITS ANNIHILATION BY
SHALLOW DONORS

   The concentration of the major electron trap (0.82eV below the
conduction band) in GaAs. (Bridgman grown) was found to increase
with increasing As pressure during growth. It was further found
that (for a given As pressure) the concentration of this trap
decreased with increasing concentration of shallow donor dopants
(Si, Se and Te). Donor concentrations above a threshold of about
$10^{17}cm^{-3}$ led to the rapid elimination of the trap. On the basis of
these findings, the 0.82eV trap was attributed to the antisite
defect As_{Ga} formed during the postgrowth cooling of the crystals.

Reprinted with permission from Appl.Phys.Lett. 1982, 40(4):342,
published by the American Institute of Physics, New York.

Lagowski, J., Lin, D.G., Aoyama, T. and Gatos, H.C.

IDENTIFICATION OF OXYGEN-RELATED MIDGAP LEVEL IN GaAs

An oxygen-related deep level ELO was identified in GaAs
employing Bridgman-grown crystals with controlled oxygen doping.
The activation energy (825±5meV) of ELO is almost the same as that
of the dominant midgap level: EL2 (815±2meV). This fact impedes the
identification of ELO by standard deep level transient
spectroscopy. However, we found that the electron capture cross
section of ELO is about four times greater than that of EL2. This
characteristic served as the basis for the separation and
quantitative investigation of ELO employing detailed capacitance
transient measurements in conjunction with reference measurements on
crystals grown without oxygen doping and containing only EL2.

Reprinted with permission from Appl.Phys.Lett. 1984, 44(3):336,
published by the American Institute of Physics, New York.

Langlois, W.E.

COMPUTER SIMULATION OF CZOCHRALSKI MELT CONVECTION IN A MAGNETIC
FIELD

The practical value of growing silicon crystals in a large
magnetic field has recently been demonstrated, and computational
investigations of the hydromagnetic melt flow have begun to appear.
These investigations, and some current work, are reviewed here.

Reprinted with permission from J.Cryst.Growth 1984, 70:73, published
by North Holland Publishing Co., Amsterdam.

Larsen, P.K., Neave, J.H., van der Veen, J.F., Dobson, P.J. and
Joyce, B.A.

GaAs(001)-c(4x4): A CHEMISORBED STRUCTURE

We have carried out an experimental study of the As-stable
GaAs(001)-c(4x4) reconstruction using a combination of
molecular-beam epitaxy with reflection electron diffraction,
angle-resolved photoemission, and surface-sensitive core-level
photoemission. (The measurements were performed at Laboratoire pour
l'Utilisation du Rayonnement Electromagnetique, Universite
Paris-Sud, F-91405 Orsay, France.) Apart from the surface symmetry,
the electron-diffraction data show one-dimensional disorder along
the [110] direction, probably corresponding to a corrugated-sheet
structure in reciprocal space. Photoemission from surface states
having dangling-bond character shows a onefold periodicity and an

energy dispersion of 0.65eV along [110], while along [110] the
dispersion is much smaller, but there is a doubling of the
periodicity. Observations of the As 3d and Ga 3d core levels show a
surface As 3d component at 0.62eV higher binding energy than the
bulk component, indicative of As-As bonding. The Ga 3d surface
core-level shift is very small and the line shape is similar to that
from the (2x4) surface. The experimental results can be understood
on the basis of an As overlayer structure, for which we propose
specific models derived from trigonally bonded chemisorbed As.

Reprinted with permission from Phys.Rev. 1983, B27:4966, published
by the American Physical Society, New York.

Lee, H.H.

A DIMENSIONLESS NUMBER FOR CVD PROCESSES: LPCVD VERSUS HPCVD

        A dimensionless number is derived which represents the extent
to which diffusion affects crystal growth in terms of variables that
can be manipulated for the crystal growth. A restriction here is
that the result is applicable to processes for which the
boundary-layer model is valid, as in the typical horizontal CVD
processes. The result is used to critically examine the low
pressure CVD process (LPCVD) used for uniform deposition of crystal
layers. It is shown that an LPCVD process does not always yield
uniform deposition and that a high pressure CVD (HPCVD) in which the
total pressure is higher than atmospheric pressure should be used in
some cases rather than an LPCVD process if the total pressure is to
be manipulated for the uniform deposition. Further, even when an
LPCVD process is warranted, the total pressure should be kept as
high as possible in some cases for uniform deposition within a
certain pressure range. The factor that determines the choice of
the total pressure is the apparent order of the growth rate with
respect to the total pressure, the choice being dependent on whether
or not the order is greater than one-half. No advantage can be
gained if the order is one-half. The dimensionless number can also
be used to specify growth conditions under which "intrinsic" growth
rate data can be generated.

Reprinted with permission from J.Cryst.Growth 1984, 68:698,
published by North Holland Publishing Co., Amsterdam.

Lodge, E.A., Booker, G.R., Warwick, C.A. and Brown, G.T.

TEM STUDY OF GROWN-IN DISLOCATIONS IN SEMI-INSULATING, UNDOPED LEC
GaAs

        The grown-in dislocations and associated point defects in SI,

undoped LEC GaAs are being characterised with the complementary
techniques of X-ray topography, infra-red imaging, cathodoluminescence and TEM. Conventional two-beam and weak-beam TEM
images are presented of dislocations in different crystals which are
decorated with precipitate particles of 15-90nm diameter. A Moire
fringe analysis of one such particle is consistent with it being
elemental As.

Reprinted with permission from 'Microscopy of Semiconducting
Materials', Eds. A.G.Cullis and D.B.Holt, Inst.Phys.Conf.Series,
1985, 76:217.

Maguire, J., Newman, R.C. and Beall, R.B.

SITE SWITCHING OF SILICON IN NEUTRON IRRADIATED AND ANNEALED GALLIUM
ARSENIDE BY VACANCY MIGRATION

Silicon doped GaAs crystals grown either by the Bridgman or LEC
technique have been irradiated with fast neutrons to doses in the
range 10^{18} to 10^{19} n^0 cm^{-2} and subsequently isochronally annealed up
to 625°C. At about 350°C the concentration of Si_{As} defects
increases but subsequently falls when the temperature exceeds
500°C. The concentration of Si_{Ga} shows the opposite behaviour. The
observations are explained by an initial release of V_{As} defects into
the crystal followed by a release of V_{Ga}. EPR measurements on the
same samples show the As_{Ga} antisite spectrum and a broad resonance
near g = 2 attributed by others to a vacancy. This latter centre
anneals between 350 and 500°C but shows no changes at lower
temperatures. The results are discussed in relation to models for
the migration of vacancies.

Reprinted with permission from J.Phys.C., 1986, 19:1897

Manasevit, H.M., Golecki, I., Moudy, L.A., Yang, J.J. and Mee, J.E.

Si ON CUBIC ZIRCONIA

Epitaxial growth of single-crystal Si films has been realised
on the (100), (110) and (111) crystallographic planes of yttria-
stabilised cubic zirconia single crystals. The Si films were grown
by chemical vapor deposition, using the pyrolysis of SiH_4 at
temperatures in the range 950-1075 °C and at deposition rates of
0.08-1.2μm/min. A predeposition annealing procedure has been
developed, resulting in a quasi-stable, oxygen-deficient zirconia
surface. A model is presented to explain the dependence of oxygen
kinetics in cubic zirconia on temperature and yttria content. The
heteroepitaxial Si films have been characterised by optical and
scanning electron microscopies, reflection electron diffraction,

X-ray diffraction, Rutherford backscattering and channeling, and
surface electrical conductivity and Hall effect measurements.
Several 0.4-0.5μm thick (100)- and (110)-oriented Si films on cubic
zirconia were found to be of higher crystal quality than commercial
(100) Si on sapphire films of similar thickness.

J.Electrochem.Soc. 1983, 130:1752. Reprinted by permission of the
publishers, The Electrochemical Society Inc., Pennington, NJ, USA.

Mandel, G.

SELF-COMPENSATION LIMITED CONDUCTIVITY IN BINARY SEMICONDUCTORS. I.
THEORY

The techniques of Kroger and Vink and Brebrick are extended to
allow a calculation of the minimum extent of self-compensation by
simple vacancies or interstitial atoms in heavily doped binary
semiconductors. The resulting equations are applied to a series of
compounds, and it is found that the degree of self-compensation by
singly ionizable vacancies varies from essentially complete in KCl
(all but ~10^{-9} of the impurities compensated) to practically none
in GaAs (only $\lesssim 10^{-3}$ of the impurities compensated). The II-VI
compounds occupy an intermediate position with about ~99 and 99.9%
self-compensation in CdTe and ZnTe, respectively. These theoretical
conductivity limitations are not sufficient to account for the
experimental limitations found in, for example, n-ZnTe or p-CdS.
The above results are extended to include multiply ionizable
vacancies, the ionization levels of which fall within the bandgap.
It is found that essentially complete self-compensation by a
combination of singly and doubly ionized vacancies will occur in the
higher bandgap II-VI compounds. As a consequence, for example, the
Fermi level in ZnTe cannot be pushed closer to the bottom of the
conduction band than half the energy separation between the second
ionization level of the acceptor vacancy and the bottom of the
conduction band. Some specific implications of the above
calculations with respect to CdTe and GaAs are discussed. Finally,
certain solubility effects (of impurities) related to stoichiometry
and the above calculations are discussed.

Reprinted with permission from Phys.Rev.1964, 134(4A):A1073,
published by the American Physical Society, New York.

Manke, C.W. and Donaghey, L.F.

ANALYSIS OF TRANSPORT PROCESSES IN VERTICAL CYLINDER EPITAXY REACTORS

Momentum, heat and mass transfer processes were studied in a
vertical cylinder reactor for the epitaxial growth of Si from SiCl

in H_2 by chemical vapor deposition. An analytical solution to the problems of heat and mass transfer in a tapered annulus is presented based on constant transport properties and fully developed laminar flow. The mean gas-phase temperature and deposition rate distribution of silicon are calculated within the reactor using a developing temperature model. Results of experimental studies of silicon deposition from $SiCl_4$ in H_2 at 1200°C in a vertical cylinder reactor are compared with the analytical results and with other models of diffusion-controlled chemical vapor deposition. This study provides an analytical method for calculating epitaxial deposition rate distributions in vertical cylinder reactors, and for designing reactors to improve the yield and uniformity of epitaxial growth.

J.Electrochem.Soc. 1977, 124:561. Reprinted by permission of the publishers, The Electrochemical Society Inc., Pennington, NJ USA.

Martin, G.M.

OPTICAL ASSESSMENT OF THE MAIN ELECTRON TRAP IN BULK SEMI-INSULATING GaAs

Near-infrared optical absorption in undoped bulk GaAs ingots is shown to be due essentially to the presence of the main deep donor EL2. Measurement of the corresponding absorption represents the first known method of quantitative determination of that level in semi-insulating material. Furthermore, complete quenching of the corresponding absorption is shown to occur under high intensity illumination. This strong effect, reported for the first time, can be directly related to the existence of a metastable state for the level EL2 presenting a strong lattice relaxation.

Reprinted with permission from Appl.Phys.Lett. 1981, 39(9):747, published by the American Institute of Physics, New York.

Masselink, W.T., Fischer, R., Klem, J., Henderson, T. and Morkoc, H.

GaAs/AlGaAs MULTIQUANTUM WELLS GROWN ON NONPOLAR SEMICONDUCTOR SUBSTRATES

We report the first study of single-crystalline GaAs grown by molecular beam epitaxy (MBE) on a nonpolar substrate. In this study, GaAs was grown under various conditions on Ge(100). Surface morphologies were specular and quite strongly dependent on surface preparation. Photoluminescence spectra from such GaAs is quite intense and is dominated by donor-acceptor recombination at 1.479eV which appears to be Ge^{o}_{Ga} - Ge^{o}_{As}. We also observe its phonon replica at 1.442eV and the recombination of bound excitons at

1.514eV. This and capacitance-voltage measurements indicate
incorporation of Ge from the substrate into the GaAs. Photoluminescence from GaAs/AlGaAs multiple quantum well structures
originates from the same processes as in bulk GaAs; in quantum
wells, however, the transitions occur at higher energies due to the
enhancement of the GaAs band gap. The observed energies are in good
agreement with theoretical predictions. We have also studied the
growth of GaAs on Si(100) and on Ge films on Si(100).

Reprinted with permission from J.Vac.Sci.Technol. 1985, B3(2):548,
published by the American Institute of Physics © 1985, American
Vacuum Society.

Massies, J., Etienne, P., Dezaly, E. and Linh, N.T.

STOICHIOMETRY EFFECTS ON SURFACE PROPERTIES OF GaAs{100} GROWN IN
SITU BY MBE

      In this paper, we report results dealing with the effects of
stoichiometry on surface properties of GaAs(001) layers grown by
MBE. Three aspects of surface properties were investigated:
crystallography, electronic properties and chemical reactivity.
Surface crystallography was studied mainly by LEED. The
reconstruction of the surface was found to be drastically dependent
on the composition of the uppermost atomic layer, ie the surface
stoichiometry. According to the arsenic surface coverage, many
structures from the c(8x2)Ga rich to the (1x1) arsenic saturated
surfaceshave been observed. The influence of stoichiometry on
surface electronic properties has been studied by electron loss
spectroscopy (ELS) and contact potential difference (CPD)
measurements. In the electron loss spectra, two peaks, at about
10.3 and 20.2 eV are very sensitive to the surface composition: they
gradually disappear when the arsenic coverage increases, and
consequently are associated with surface states on gallium atoms.
On the other hand, the CPD measurements have shown that the
variation of the work function with the arsenic surface coverage is
not monotonic: in particular, an abrupt change of work function of
about 300meV occurs between the (1x6) and c(2x8) structures which
are very similar as far as the arsenic surface coverage (about 0,5
and 0,6 respectively) is concerned. Therefore, it seems that the
work function is strongly dependent on the atomic reconstruction
occurring at the surface, and not only on its stoichiometry. The
connection between stoichiometry and chemical reactivity of the
surface is illustrated by the study of H_2S adsorption: a large
difference (factor of 10^3) in sticking coefficient has been found
between surfaces with different arsenic coverages.

Reprinted with permission from Surf.Sci. 1980, 99:121, published by
Elsevier Science Publishing Co., Amsterdam.

Matsuo, S. and Kiuchi, M.

LOW TEMPERATURE CHEMICAL VAPOR DEPOSITION METHOD UTILISING AN
ELECTRON CYCLOTRON RESONANCE PLASMA

The plasma deposition apparatus developed in this study can
realise a deposition of dense and high quality thin films, such as
Si_3N_4 and SiO_2, without the need for substrate heating. It does
this by enhancing the plasma excitation efficiency at low gas
pressures (10^{-4} Torr) and by the acceleration effect of ions with
moderate energies (10 to 20 eV), using a microwave ECR (Electron
Cyclotron Resonance) excited plasma, and a plasma stream extraction
onto the specimen table by a divergent magnetic field method. The
Si_3N_4 and SiO_2 films deposited are comparable to those prepared by
high temperature CVD and thermal oxidation, respectively, in
evaluations such as by buffered HF solution etch rate measurement.

Reprinted with permission from Jpn.J.Appl.Phys. 1983, 22(4):L210,

Mihelčić, M. and Wingerath, K.

NUMERICAL SIMULATIONS OF THE CZOCHRALSKI BULK FLOW IN AN AXIAL
MAGNETIC FIELD: EFFECTS ON THE FLOW AND TEMPERATURE OSCILLATIONS IN
THE MELT

The influence of an external axi-symmetric magnetic field of
various intensities on the flow and temperature field in a model of
the Czochralski crystal growth process was simulated. It was found
that not only the flow velocity vectors are damped, but also the
strong flow and temperature oscillations are eliminated. It appears
that an external magnetic field of 2000 G (for a large-scale model,
see table 1) will suffice in order to stabilise both the flow and
temperature field.

Reprinted with permission from J.Cryst.Growth 1985, 71:163,
published by North Holland Publishing Co., Amsterdam.

Mirsky, U. and Shechtman, D.

TRANSMISSION ELECTRON MICROSCOPY AND X-RAY TOPOGRAPHY STUDY OF
CADMIUM MERCURY TELLURIDE

The fault structure of MCT crystals was studied by means of
X-ray topography and transmission electron microscopy. Subgrain
boundaries were revealed and identified by both techniques and their
tilt angle was calculated to be of the order of 10^{-3} rad. The
dislocations which form the subgrain boundaries were identified by
TEM to have Burger's vectors parallel to <110> directions. Twins in

large numbers were observed by TEM. The twins were identified to
have a {111} habit plane.

Reprinted with permission from J.Electron.Mater. 1980, $\underline{9}$(6):933,
published by The Metallurgical Society of AIME, Warrendale, PA, USA.

Mönch, W.

SOME ASPECTS OF SURFACE SCIENCE RELATED TO MBE

Crystal growth by the technique of molecular beam epitaxy
occurs on clean surfaces under well controlled ultra-high vacuum
conditions. The quality of the growing film is usually monitored *in
situ* by the observation of surface-sensitive electron diffraction.
Therefore, knowledge about the properties of clean surfaces is
essential for MBE. This contribution will consider GaAs only, since
with surfaces of this semiconductor a wealth of experimental data is
available. In particular, we will discuss the crystallography of
(111) and (001) surfaces, which are used in MBE, and the correlation
between different reconstructions and the respective chemical
surface compositions. Another issue is occupied and empty intrinsic
surface states as well as extrinsic electronic surface properties.
Finally, results on the chemisorption of oxygen, of hydrogen, and of
H_2S are examined. With all the different topics the properties of
cleaved surfaces are looked at for reasons of comparison. Studies
with such surfaces have recently questioned the bulk homogeneity of
the substrates used in MBE since segregation of anion atoms was
detected on surfaces cleaved from GaP, GaAs, GaSb and InP.

Reprinted with permission from "MBE and Heterostructures", Eds.
L.L.Chang and K.Ploog, NATO ASI Series E - Applied Sciences No 87,
1985. Martinus Nijhoff, Dordrecht, Netherlands.

Moss, R.H.

ADDUCTS IN MOVPE OF III-V COMPOUNDS

Metalorganic vapour phase epitaxy has advanced rapidly in
recent years and the success of the technique for GaAs and GaAlAs is
well known. Its extension to other III-V materials has met with
problems due to instabilities of the alkyl starting materials and
unwanted side reactions in the gas phase. These problems are
particularly noted in the growth of indium containing compounds.
One technique which has been successfully used to overcome these
problems employs Lewis acid-base adducts of group III alkyls and
group V alkyls [1]. These adducts may be formed in situ in the gas
phase by mixing the corresponding alkyls: eg: In (CH_3) + $P(C_2H_5)_3$ $\rightarrow$
$(CH_3)_3$ $\leftarrow$ $P(C_2H_5)_3$. Alternatively, the adducts may be prepared by

standard chemical techniques and used as source materials in MOVPE
systems. The adducts being thermally more stable than alkyls are
safer to handle and offer advantages over conventional alkyl
materials.

The two methods of using adducts in MOVPE will be described and
results will be presented for the growth of InP,(Ga,In)As and
(Ga,In)(As,P). The results of p-type doping of InP with Zn will be
reported along with present device applications of the growth
technique, particularly to the overgrowth of gratings for
Distributed Feedback (DFB) lasers.

The merits of the adduct techniques will be critically assessed
in the light of the present results and future developments
discussed.

Reprinted with permission from J.Cryst.Growth 1983, 65:463,
published by North Holland Publishing Co., Amsterdam.

Müller-Krumbhaar, H.

MASTER-EQUATION APPROACH TO STOCHASTIC MODELS OF CRYSTAL GROWTH

A master-equation approach is formulated to obtain kinetic
equations for stochastic models of the crystal-vapor interface. The
properties of two Ising-type models are studied in stable and
metastable states. General transition probabilities for the
adsorption and evaporation of atoms at the interface are introduced,
which may account for different types of dynamic behavior. Marked
dependence of the interface kinetics upon the details of the
transition probabilities is found, in contrast to the case of
homogeneous systems.

Reprinted with permission from Phys.Rev. B10(4):1308, 1974,
published by the American Physical Society, New York.

Mullin, J.B., Irvine, S.J.C. and Ashen, D.J.

ORGANOMETALLIC GROWTH OF II-VI COMPOUNDS

Developments in the organometallic growth of II-VI compounds
are reviewed with an emphasis on epitaxial growth related problems.
The advantages and disadvantages of organometallic growth are
compared with analogous growth using conventional VPE techniques.
Consideration is given to the growth of specific compounds using
alkyls of Zn, Cd or Hg with the elements O, S, Se and Te as hydrides
or alkyls. Special reference is given to the decomposition
characteristics of typical alkyls such as (C_2H_5) Te, $(CH_3)_2Te$ and

$(CH_3)_2Cd$ by pyrolysis in H_2. Reaction mechanisms involving homogeneous and heterogeneous pyrolysis, adduct formation and the potential role of adsorption on parameters such as growth rate are discussed.

Reprinted with permission from J.Cryst.Growth 1981, 55:92, published by North Holland Publishing Co., Amsterdam.

Mullin, J.B. and Irvine, S.J.C.

VAPOUR PHASE EPITAXY OF $Cd_xHg_{1-x}Te$ USING ORGANOMETALLICS

A new open-flow vapour phase process is reported for the epitaxial growth of cadmium mercury telluride on to substrates of cadmium telluride. The layers have been characterised by in-depth profiling using secondary ion mass spectrometry and also the electrical properties have been assessed using the van der Pauw technique. Material with x values in $Cd_xHg_{1-x}Te$ ranging from 0 to 0.5 have been grown at ~410°C.

Reprinted with permission from J.Phys.D. 1981, 14:L149, published by the Institute of Physics, Bristol, UK.

Mullin, J.B., Irvine, S.J.C., Royle, A., Tunnicliffe, J., Blackmore, G. and Holland, R.

A STUDY ON THE PURITY AND SIMS PROFILING OF MOVPE-CADMIUM MERCURY TELLURIDE

A preliminary study has been carried out on the electrical properties and chemical purity of epitaxial layers of cadmium mercury telluride (CMT) grown by metal-organic vapor phase epitaxy (MOVPE). It has been shown that the mobility of such layers grown at 410°C are comparable with bulk CMT of equivalent carrier concentration. SIMS profile studies of trace impurities in the epitaxial layers and CdTe substrates used for epitaxial deposition revealed significant impurity concentration enhancement (ICE effect) for a number of elements (Li, Na, K, Al, Ga, In and Mn) at the CdTe/CMT interface; some of these elements (Li, Na, K, Al and In) also showed this ice effect at the subsequent CMT/CdTe interface. The magnitude of the ICE effect could be quite large (a factor of ~100 for Li) but its origin was not identified although potential mechanisms for its formation are discussed.

Reprinted with permission from J.Vac.Sci.Technol. 1983, A1(3):1612, published by the American Institute of Physics © 1983, American Vacuum Society.

Nagao, S., Higashitani, K., Akasaka, Y. and Nakata, H.

CRYSTALLINE FILM QUALITY IN REDUCED PRESSURE SILICON EPITAXY AT LOW TEMPERATURE

The influence of deposition pressure on epitaxial crystalline film quality has been investigated with respect to vapor-phase silicon epitaxial growth (SiH_2Cl_2 - H_2 system). It is shown that the crystalline film quality has been improved by reducing the deposition pressure. In addition, a prebaking process at reduced pressure has been found to be effective for obtaining single-crystalline silicon films with atmospheric pressure deposition as well as reduced pressure deposition. The deposition temperature can be lowered down to 930°C with perfect crystalline film by reducing the deposition pressure. The crystalline quality of epitaxial films grown at a low temperature of 930°C using the reduced pressure technique has been verified to be excellent by fabricating bipolar transistors with an oxide isolation technique.

Reprinted with permission from J.Appl.Phys. 1985, 57(10):4589, published by the American Institute of Physics.

Neave, J.H. and Joyce, B.A.

STRUCTURE AND STOICHIOMETRY OF (100) GaAs SURFACES DURING MOLECULAR BEAM EPITAXY

Relationships between surface structures on (100) GaAs substrates and incident fluxes of Ga and As have been examined in detail over the temperature range 300-900 K. Surface reconstruction effects observed by RHEED have been related to surface stoichiometry, and the stability criteria for several reconstructed surfaces determined. Reconstruction observed during auto-epitaxial growth of GaAs from beams of As_4 and Ga is not a simple function of the incident fluxes, since effects associated with the thermal stability of the substrate can have important consequences. The results are discussed in relation to surface kinetic information obtained from modulated beam measurements, and to some current theories of surface reconstruction and growth step behaviour. Results on surface topography are used to illustrate the ambiguities of streaked RHEED patterns. It is shown that streaks of constant intensity normal to the shadow edge cannot be used as the sole criterion by which to define the presence of a flat surface.

Reprinted with permission from J.Cryst.Growth 1978, 44:387, published by North Holland Publishing Co., Amsterdam.

Nishitani, K., Ohkata, R. and Murotani, T.

MOLECULAR BEAM EPITAXY OF CdTe and $Hg_xHg_{1-x}Te$ ON GaAs(100)

Using the molecular beam epitaxial (MBE) technique, CdTe and $Hg_{1-x}Cd_xTe$ have been grown on Cr-doped GaAs (100) substrates. A single effusion cell charged with polycrystalline CdTe is used for the growth of CdTe films. The CdTe films grown at 200 C with a growth rate of ~2 μm/hr show both streaked and "Kikuchi" patterns, indicating single crystalline CdTe films are smoothly grown on the GaAs substrates. A sharp emission peak is observed at near band-edge (7865A, 1.577 eV) in the photoluminescence spectrum at 77 K. For the growth of $Hg_{1-x}Cd_xTe$ films, separate sources of HgTe, Cd and Te are used. $Hg_{0.6}Cd_{0.4}Te$ films are grown at 50°C with a growth rate of 1.7 μm/hr. The surfaces are mirror-smooth and the interfaces between the films and the substrates are very flat and smooth. As-grown $Hg_{0.6}Cd_{0.4}Te$ films are p-type and converted into n-type by anealing in Hg pressure. Carrier concentration and Hall mobility of an annealed $Hg_{0.6}Cd_{0.4}Te$ film are 1 x 10^{17} cm^{-3} and 1000 cm^2/V-sec at 77 K, respectively.

Reprinted with permission from J.Electron.Mater. 1983, 12(3):619, published by the Metallurgical Society of AIME, Warrendale, PA, USA.

Noreika, A.J., Farrow, R.F.C., Shirland, F.A., Takei, W.J., Greggi Jnr., J., Wood, S. and Choyke, W.J.

CHARACTERISATION OF MOLECULAR BEAM EPITAXIALLY GROWN HgCdTe ON CdTe AND InSb BUFFER LAYERS

Molecular beam epitaxially (MBE) grown layers of (001)CdTe and (001)HgCdTe were examined by RHEED, DCRC, XTEM and EDS to assess the effect of growth parameters on crystal perfection. The CdTe layers were further investigated via low temperature photoluminescence (PL) and SIMS analysis to determine potential impurity incorporation from InSb substrates. In both sets of analysis substrate temperatures and substrate preparation were the parameters which most influenced film growth. Double crystal X-ray rocking curve (DCRC) data showed weak substrate dependence in lattice matched (001)CdTe. PL data showed a stronger influence of substrate temperature on radiative defect incorporation. The HgCdTe alloy also showed structural dependence on substrate temperature in XTEM micrographs of precipitate formation which were confirmed by *in situ* EDS. Lateral compositional uniformity of alloy layers was determined by step-and-repeat infrared transmission measurements across a 3 in. GaAs wafer. The data have indicated an average 300 K cutoff wavelength of 7.05 ± 0.5 μm across a full diameter.

Reprinted with permission from J.Vac.Sci.Technol. 1986, A4(4):2081,

Normand, C., Pomeau, Y. and Velarde, M.G.

CONVECTIVE INSTABILITY: A PHYSICIST'S APPROACH

A number of apparently disparate problems from engineering, meteorology, geophysics, fluid mechanics and applied mathematics are considered under the unifying heading of natural convection. After a review of the mathematical framework that serves to delineate these problems, the heuristic approach to Benard and Rayleigh convection is discussed with special attention to buoyancy and surface tension. Then consideration is given to some aspects of scaling, and the nondimensionalisation of equations to a given problem. The thermohydrodynamic description of a Newtonian fluid is presented, and the Boussinesq-Oberbeck model. This is followed by a treatment of the linear stability problem, and a description of the basic ideas of Landau and Hopf concerning the bifurcation of secondary solutions. Quantitative, though approximate, estimations are given for quantities belonging to the nonlinear steady convective regime: flow velocity and temperature distribution. Higher-order, though steady, bifurcations are discussed, as well as the transition to turbulence, along with such time-dependent phenomena as relaxation oscillations. The paper concludes with an Appendix showing a simple application of the Leray-Schauder topological degree of a mapping.

Reprinted with permission from Rev.Mod.Phys. 1977, 49(3):581, published by the American Physical Society, New York.

Olsen, G.H.

INTERFACIAL LATTICE MISMATCH EFFECTS IN III-V COMPOUNDS

The effects of lattice mismatch upon the microstructure, macrostructure and device performance of III-V heteroepitaxial structures have been evaluated. Lattice mismatched VPE heteroepitaxial materials are shown to exhibit misorientation, tetragonal distortion, cracking and cross-hatching, all of which can be related to the relief (or non-relief) of lattice mismatch. Materials which are grown under tension are shown to crack much more readily than those in compression, and require a significantly thicker compositionally graded region in order to completely eliminate cracking. The electro-optical properties of transmission photocathodes and transmission secondary electron multipliers are found to be very sensitive functions of lattice mismatch and correlate well with transmission electron microscope evidence from these devices.

Reprinted with permission from J.Cryst.Growth 1975, 31:223, published by North Holland Publishing Co., Amsterdam.

Olson, J.M. and Rosenberger, F.

CONVECTIVE INSTABILITIES IN A CLOSED VERTICAL CYLINDER HEATED FROM BELOW: PART I. MONOCOMPONENT GASES

     Kr, Xe and $SiCl_4$ have been investigated for convective instabilities in closed vertical cylinders with conductive walls heated from below. Critical Rayleigh numbers N^i_{Ra} for the onset of various convective modes (including the onset of marginally stable and periodic flow) have been determined with a high resolution differential temperature sensing method. Flow patterns were deduced from a multiple sensor arrangement. For the three lowest modes (i=1,2,3) good quantitative agreement with linear stability theory is found. Stable oscillatory modes (periodic fluctuations of the mean flow) with a period of approximately 5 s are found for a relatively narrow range of N_{Ra}. The critical Rayleigh number $N_{Ra}(OSC)$ for the onset of oscillatory temperature fluctuations is 134 ± 50 for an aspect ratio (height/radius) of 6.

Reprinted with permission from J.Fluid Mech. 1979, 92(4):609, published by Cambridge University Press, Cambridge, UK.

Olson, J.M. and Rosenberger, F.

CONVECTIVE INSTABILITIES IN A CLOSED VERTICAL CYLINDER HEATED FROM BELOW.   PART 2.   BINARY GAS MIXTURES

     Non-reactive binary gas mixtures (Xe+He, $SiCl_4 + H_2$) have been investigated for convective instabilities in closed vertical cylinders with conductive walls heated from below. Critical Rayleigh numbers N^i_{Ra} for the onset of various convective modes (including the onset of marginally stable and periodic flow) have been determined with a high resolution differential temperature sensing method. It is found that the second component can significantly alter the hydrodynamic state of the fluid compared to the monocomponent behaviour. Considerably lower critical *thermal* Rayleigh numbers for steady and time dependent convective modes are observed. The Xe:He system shows stable oscillatory modes similar to those observed in monocomponent gases (periodic disturbances of the mean flow, $T_0 \approx 5s$) from N_{Ra} = 713 to 780, where a new mode with T_0 = 15s sets in. The frequency of these slower temperature oscillations can be fitted by an equation of the form $f^2 = k' (N_{Ra} - N^0_{Ra})$ where k' and N^0_{Ra} are constants, which supports the contention that these oscillations are the result of vertical vorticity. For $SiCl_4:H_2$ the high frequency oscillations occur only as a transient mode eventually evolving into the low frequency mode characteristic of binary gas mixtures. This low frequency state is degenerate with a stable time-independent state over a considerable range of N_{Ra}. Finite amplitude perturbations can lead to (1) transient oscillatory phenomena accompanied by reorientations of the roll cells with mean periods of 3-5 min; and (2) stable oscillatory flow at N_{Ra}'s considerably below $N_{Ra}(OSC)$. The unique behaviour of these binary fluids is tentatively assigned to thermal diffusion.

Reprinted with permission from J.Fluid Mech. 1979, 92(4):609, published by Cambridge University Press, Cambridge, UK.

Osgood Jr, R.M. and Ehrlich, D.J.

OPTICALLY INDUCED MICROSTRUCTURES IN LASER-PHOTODEPOSITED METAL FILMS

Regular 0.15 to 0.5 μm-period ripple structures have been seen in laser-photodeposited Cd and Zn films. Electron-microscope observations have established the evolution of these microstructures in the growing films.

Reprinted with permission from Optics Letters 1982, 7(8):385, published by the Optical Society of America.

Pande, K.P. and Seabaugh, A.C.

LOW TEMPERATURE PLASMA-ENHANCED EPITAXY OF GaAs

Low temperature ($\leqslant 450°C$) deposition of single-crystal GaAs using a new plasma-enhanced MO-CVD technique is reported. In this technique, plasma is created by a dc potential and the substrate is not directly exposed to the plasma. Deposition of GaAs was achieved at extremely low plasma power (0.3-0.5 W cm^2) using trimethyl-gallium (TMGa) and arsine (or trimethylarsenic) reactants. The resulting epitaxial films show excellent surface morphology and thickness uniformity over a large area substrate. A linear dependence of growth rate upon TMGa concentration was observed with a typical growth rate of 0.1 μm/min for a TMGa flow rate of 15 ml/min. Undoped films were found to be n-type with a room temperature mobility in the range of 5200 cm^2 V^{-1}.s^{-1}. Measurements on Schottky barrier devices fabricated on n/n+ layers show uniform impurity doping profiles. Temperature dependence of the diode capacitance indicates a density of deep trapping centers as low as 6.2 x 10^{13} cm^{-3}.

J.Electrochem.Soc. 1984, 131:1357. Reprinted by permission of the

Panish, M.B.

MOLECULAR BEAM EPITAXY OF GaAs AND InP WITH GAS SOURCES FOR As AND P

Arsine and phosphine were decomposed in a high temperature
leak-source to provide As_2 and P_2 molecular beams for molecular beam
epitaxy of GaAs and InP. Reasonable growth rates (0.5-1.5 μm/hr)
were achieved for both semiconductors. The studies demonstrated
that with this method, MBE of GaAs can be done with the substrate
temperature as high as 700°C. The maximum growth temperature
obtained for InP was approximately 600°C. A reasonable increase in
leak rate at the source should permit GaAs-MBE at 750°C and InP-MBE
at 650°C.

Panish, M.B., Temkin, H. and Sumski, S.

GAS SOURCE MBE OF InP AND $Ga_xIn_{1-x}P_yAs_{1-y}$: MATERIALS PROPERTIES AND
HETEROSTRUCTURE LASERS

The growth of useful $Ga_xInP_yAs_{1-y}$ by MBE is inhibited by the
difficulty in maintaining precise control over the flux from
condensed source As and P effusion ovens. The use of P_2 and As_2
beams generated from the cracking of PH_3 and AsH_3 as an alternative
approach is reviewed and new data are presented. Accommodation
coefficients were determined for P_2 and As_2 on heated InP and GaAs
surfaces. The flux composition as a function of epitaxial layer
composition was determined for the lattice-matching quaternary.
Studies of doping were done and injection laser structures were
grown. Although detailed studies of layer composition control were
not done, the achievable control seems quite adequate for injection
lasers. A variety of heterostructure lasers emitting at nominally
1.5 μm with broad area 300-K threshold-current densities from 1300
to 2400 A/cm^2 were demonstrated. Ridge waveguide lasers fabricated
from the same wafers had I_{th} of about 40 mA and differential
efficiencies of 36%-45%. They lased cw to 50°C.

Park, R.M., Mar, H.A. and Salansky, N.M.

MOLECULAR BEAM EPITAXY GROWTH OF ZnSe ON (100)GaAs BY COMPOUND
SOURCE AND SEPARATE SOURCE EVAPORATION: A COMPARATIVE STUDY

ZnSe layers have been grown by MBE by both compound source and
separate source evaporation on (100)GaAs substrates. Photoluminescence measurements on layers grown from a compound source
showed the material to contain a high concentration of impurities
and defects as indicated by the presence of large DA pair
recombination and SA emission peaks in the PL spetra. Layers grown
from separate sources were grown at various Zn to Se beam pressure
ratios ranging from 1 to 30, and with substrate temperatures ranging
from 300 to 400°C. For layers grown at 900 A/h with a Zn to Se beam
pressure ratio of 1:1 the 4.2 K PL spectra were dominated by a
Ga-bound exciton at 2.7982 eV the intensity of which was strongly
substrate temperature dependent. It was found that the Ga-bound
exciton peak could be suppressed by using higher Zn to Se ratios,
and at beam pressure ratios >10 the dominant exciton peak was a
free-exciton related peak at 2.800 eV. DA pair recombination was
not detectable in layers grown from separate sources. Layers grown
from separate sources at a higher growth rate (~0.4/h) contained
additional defects indicated by the presence of an I_x peak (exciton
bound to a native defect) and a much smaller unidentified peak at
2.49 eV.

Reprinted with permission from J.Vac.Sci.Technol. 1985, B3(2):676,
published by the American Institute of Physics, New York, © 1985
American Vacuum Society.

Pautrat, J.L., Magnea, N. and Faurie, J.P.

THE SEGREGATION OF IMPURITIES AND THE SELF-COMPENSATION PROBLEM IN
II-VI COMPOUNDS

The electrical and optical properties of ZnTe or CdTe crystals
are generally inhomogeneous after annealing. This is shown to be
due to the redistribution of preexisting impurities. Indeed, after
growth, these crystals contain some excess tellurium which must
precipitate since the solidus line has a retrograde shape. The
liquid Te droplets are then able to purify the surrounding materials
by a solid-liquid segregation mechanism. On annealing, the
tellurium excess can disappear more or less rapidly: the Te
precipitates and the small inclusions shrink progressively while the
impurities are released into the crystal. This mechanism explains
the bright dots observed in the scanning electron microscope
(cathodoluminescence mode) and due to the high lithium
concentration. Under other conditions, dark dots are obtained and
the role of copper is suspected. This model is discussed on the

basis of the available data (segregation and diffusion coefficients). Further confirmation is obtained from direct chemical analysis of the crystal and of the inclusions itself. The consequences of the segregation of impurities on the interpretation of the self-compensation problem are analysed. A fully satisfactory interpretation of conductivity variation with zinc pressure previously attributed to the zinc vacancy, can be derived with the single hypothesis that acceptor and donor impurities in sufficient amount are stored in the crystal. The dominant role of residual impurities in the close control of the II-VI materials is particularly emphasised.

Reprinted with permission from J.Appl.Phys. 1982, $\underline{53}$(12):8668, published by the American Institute of Physics, New York.

Penning, P.

GENERATION OF IMPERFECTIONS IN GERMANIUM CRYSTALS BY THERMAL STRAIN

Thermal strain is induced in a crystal if it is heated inhomogeneously. These strains can be so large that either plastic flow or cracking may take place. Both effects have been observed in crystals quenched from high temperature. In the case of plastic flow the distribution of dislocations, as revealed by etching, shows marked characteristics, dependent on the way of quenching and the orientation of the crystal. It is shown that these characteristics can be explained very well if it is assumed that the amount of plastic flow is directly proportional to the elastic strain induced by the non-uniform temperature. During the growth of a crystal from a melt, thermal strain may also very well be present in the crystal The main sources of these strains are indicated and their influence on the perfection of the growing crystal are discussed. Secondary effects of plastic flow due to thermal strain in grown crystals and quenched samples, such as residual stresses and the freezing in of defects, are discussed briefly.

Reprinted with permission from Philips Research Repts., 1958, $\underline{13}$:79.

Pessa, M., Huttunen, P. and Herman, M.A.

ATOMIC LAYER EPITAXY AND CHARACTERISATION OF CdTe FILMS GROWN ON CdTe(110) SUBSTRATES

It is shown that growth of homoepitaxial CdTe films on CdTe(110) substrates by the atomic layer epitaxy method can lead to high structural perfection. The electronic structure of these films with the thickness of approximately 100A was studied at room temperature by means of angle-resolved photoemission. In addition to several bulklike features observed in the spectra an intrinsic surface state is believed to be found at the energy of 0.33 eV below the valence-band maximum for the first time in epitaxially grown II-VI compound semiconductor films. The work function of the films was found to be 5.75 eV ± 0.05 eV independent of whether the growth was terminated by a Cd pulse or a Te_2 pulse.

Reprinted with permission from J.Appl.Phys. 1983, 54(10):6047, published by The American Institute of Physics, New York.

Phillips, J.M. and Gibson, J.M.

THE GROWTH AND CHARACTERISATION OF EPITAXIAL FLUORIDE FILMS ON SEMICONDUCTORS

This review considers the growth and characterisation of epitaxial alkaline earth fluoride compounds on semiconductors. The field has developed quite rapidly in recent years, from the original demonstration of epitaxial growth in these systems to the investigation of the structure and properties of the layers. Two recent developments are reviewed in detail: the epitaxial relations which have been discovered between fluoride films and semiconductor substrates and the variety of interface structures which these systems display. These findings are beginning to lead to an understanding of the epitaxial growth process in these systems and how it relates to that in other more thoroughly studied systems.

Reprinted with permission from "Thin Films and Interfaces II", Eds. J.Baglin, D.R.Campbell and W.K.Chu, MRS Symp.Proc. 25:381. Copyright 1985, Elsevier Science Publishing Co.Inc.

Ploog, K.

MOLECULAR BEAM EPITAXY OF III-V COMPOUNDS

An important aspect of the large expansion in the development and production of solid-state devices has been the increasing demand for more sophisticated multilayered semiconductor crystals. During the last decade a new thin film growth technique, molecular beam epitaxy (MBE) has been developed to provide improved control over composition, thickness and doping profile in the direction of growth on an atomic scale. Based on these unique features, molecular beam epitaxy offers the feasibility of tailoring the electronic properties of the crystal to its desired function by growing high-quality epitaxial thin films with predetermined compositional and/or doping profiles perpendicular to the growth surface. The attractive capabilities of MBE for large-area uniform growth with extreme control over microscopic dimensions suggest that both

characteristic and reproducibility of microwave and
optoelectronicdevice structures may be improved. It is the purpose
of this article to review the basic process of molecular beam
epitaxy as a crystal growth technique, to indicate present trends
and material problems, and - at a time of continuing innovation - to
show promise for future application. The emphasis of this review is
on the growth and properties of crystalline III-V compound
semiconductor thin films focusing mainly on GaAs and $Al_xGa_{1-x}As$.

'Molecular Beam Epitaxy' in "Crystals, Growth Properties and
Applications", $\underline{3}$:73, published by Springer Verlag, Heidelberg, 1980.

Ponce, F.A.., Yamashita, T. and Hahn, S.

STRUCTURE OF THERMALLY INDUCED MICRODEFECTS IN CZOCHRALSKI SILICON
AFTER HIGH-TEMPERATURE ANNEALING

The microstructure of Czochralski silicon annealed at high
temperaturs has been investigated using high resolution transmission
electron micoscopy. After prolonged heat treatments at temperatures
close to 1200°C, two types of microdefects are observed: (a) small
polyhedral oxide precipitates, with typical diameters of about 15 nm
and facets along crystalline plane3s of the silicon matrix, and (b)
planar faults along {111} planes which have been directly identified
as extrinsic stacking faults bounded by Frank-type dislocations.

Reprinted with permission from Appl.Phys.Lett. 1983, $\underline{43}$:1051,
published by the American Institute of Physics, New York.

Ponce, F.A.

STRUCTURE OF MICRODEFECTS IN SEMICONDUCTING MATERIALS

Microdefects are common in semiconducting materials of the
highest quality. Some are present in as-grown materials, others
appear during subsequent processing, especially when subjected to
thermal treatments. Their identification has not been possible
until very recently because their size is relatively too small to be
observed by conventional electron microscopy. A review of recent
applications of HREM to the study of microdefects in silicon and
GaAs is presented in this paper.

Reprinted with permission from 'Microscopy of Semi-conducting
Materials', Eds. A.G.Cullis and D.B.Holt, Inst. of Physics Conf.Ser.
$\underline{76}$:1, Institute of Physics, Bristol, UK, 1985.

Pond, R.C., Gowers, J.P., Holt, D.B., Joyce, B.A., Neave, J.H. and Larsen, P.K.

A GENERAL TREATMENT OF ANTIPHASE DOMAIN FORMATION AND IDENTIFICATION AT POLAR-NONPOLAR SEMICONDUCTOR INTERFACES

Observations of antiphase disorder obtained using a new technique of transmission electron microscopy, in (100) epitaxial layers of GaAs:Ge produced by MBE are presented. The crystallographic origin of this type of disorder is analysed using a recently developed approach. It is shown that anti-phase disorder can exist in epitaxial layers where the substrate orientation is of the form (hk0); this is consistent with experimental observations that disorder is observed on (100) and (110) substrates, but not on (111) or (211), for example. Antiphase boundaries in (hk0) specimens are shown to separate interfacial domains which are energetically degenerate and related by symmetry operations.

van der Putte, P., Giling, L.J. and Bloem, J.

GROWTH AND ETCHING OF SILICON IN CHEMICAL VAPOUR DEPOSITION SYSTEMS; THE INFLUENCE OF THERMAL DIFFUSION AND TEMPERATURE GRADIENT

In the growth of epitaxial silicon using $SiCl_4$ as source material it is known that the growth rate at first increases with increasing p_{SiCl_4}, but at higher p_{SiCl_4} decreases and finally becomes negative. The exact shape of the growth curve and especially the point of zero growth, where the etch rate counterbalances the growth rate, differs considerably amongst various investigators. This phenomenon was subject to theoretical considerations. Three models were evaluated: (i) a simple equilibrium model in which gas transport limited kinetics are used with one single diffusion coefficient for all species; (ii) an extended model using individual diffusion coefficients; (iii) a model as in (ii) but including the effect of thermal diffusion. It is shown that models (i) and (ii) do not give a general solution of the problem. The best results are obtained with model (iii). The growth rate appears to be very sensitive to the temperature gradient ∇T. As ∇T depends on the kind of apparatus used in the deposition process, it provides a reasonable explanation for the observed differences.

Putz, N., Veuhoff, E., Heinecke, H., Heyen, M., Luth, H. and Balk, P.

GaAs GROWTH IN METALORGANIC MBE

    In this paper we report on the growth of GaAs in a UHV system
from molecular beams of trimethyl gallium (TMG) and arsine (AsH_3).
Deposition could only be achieved if the AsH_3 was partially
decomposed before injecting it into the system. If arsenic is
provided in excess, the growth rate depends linearly on the TMG
pressure. The growth rate saturates the system if the arsenic
pressure in the beam is reduced; the saturation value being
proportional to the arsenic flux. The results can be explained by a
model assuming that sticking of TMG at the growing surface is
allowed only if As simultaneously is supplied in excess. The layers
are heavily doped with carbon (p $\approx 10^{19}$-10^{20} cm^{-3}). The carbon
uptake is reduced by several orders of magnitude if TEG is used
rather than TMG.

Reprinted with permission from J.Vac.Sci.Technol. 1985, $\underline{B3}$(2):671,
published by the American Institute of Physics, New York, © 1985,
American Vacuum Society.

Qadri, S.B., Goldenberg, M., Prinz, G.A. and Ferrari, J.M.

X-RAY CHARACTERISATION OF SINGLE-CRYSTAL Fe FILMS ON GaAs GROWN BY
MOLECULAR BEAM EPITAXY

    X-ray structural studies of Fe films grown on (110)GaAs by MBE
have shown very well-oriented mosaic structure originating at the
interface and presumably caused by the lattice mismatch of 1.34%.
The size and structure of the domains depend upon growth conditions,
but at film thicknesses approaching 1μm, the films assume the
character of bulk α-phase Fe. This study has shown that although
RHEED photographs indicated smooth flat MBE growth of these films,
they were inadequate to explain the anomolous transport and magnetic
properties. On the other hand, careful X-ray studies, especially
linewidth rocking-curve measurements have revealed a very fine grain
mosaic structure which can strongly affect those properties.

Reprinted with permission from J.Vac.Sci.Technol. 1985, $\underline{B3}$(2):718,
published by the American Institute of Physics, New York, © 1985,
American Vacuum Society.

Queisser, H.J.

ELECTRICAL PROPERTIES OF DISLOCATIONS AND BOUNDARIES IN
SEMICONDUCTORS

Simple models have been suggested to predict electronic properties of lattice defects in semiconductor crystals: dislocations ought to act via the acceptor character of dangling bonds, and small-angle grain boundaries ought to consist of regular arrays of dislocations. The actual situation in most semiconductors is, however, much more complicated. The observed electrical effects of dislocations do not confirm the dangling-bond concept, they are affected by dissociation and reconstruction. There appear to be differences between straight and kinked dislocations. Dislocations owe much of their electronic behavior to clouds and precipitates of impurities; oxygen in silicon plays a significant role. This review summarises the present status of experimental methods and results, including luminescence and capacitance spectroscopy as well as mapping and imaging techniques using electron-microscopes.

Reprinted with permission from 'Defects in Semiconductors' 1983 $\underline{2}$:323, Eds. S.Mahagan and J.W.Corbett, published by North Holland Publishing Co., Amsterdam.

Razeghi, M., Poisson, M.A., Larivain, J.P. and Duchemin, J.P.

LOW PRESSURE METALORGANIC CHEMICAL VAPOR DEPOSITION OF InP AND RELATED COMPOUNDS

The low pressure metalorganic chemical vapor deposition epitaxial growth and characterisation of InP, $Ga_{0.47}In_{0.53}As$ and $Ga_xIn_{1-x}As_yP_{1-y}$ lattice-matched to InP substrate are described. The layers were found to have the same etch pit density (EPD) as the substrate. The best mobility obtained for InP was 5300 cm^2 V^{-1}S^{-1} at 300 K and 58 900 cm^2 V^{-1}S^{-1} at 77 K and for GaInAs was 11900 cm^2 V^{-1}S^{-1} at 300 K, 54 600 cm^2 V^{-1}S^{-1} at 77 K and 90 000 cm^2 V^{-1}S^{-1} at 2°K. We report the first successful growth of a GaInAs-InP super-lattice and the enhanced mobility of a two dimensional electron gas at a GaInAs - InP heterojunction grown by LP-MO CVD. LP MO CVD material has been used for GaInAsP-InP, DH lasers emitting at 1.3 μm and 1.5 μm. These devices exhibit a low threshold current, a slightly higher than liquid phase epitaxy devices and a high differential quantum efficiency of 60%. Fundamental transverse mode oscillation has been achieved up to a power output of 10 mW. Threshold currents as low as 200 mA dc have been measured for devices with a stripe width of 9 μm and a cavity length of 300 μm for emission at 1.5 μm. Values of T_0 in the range of 64-80°C have been obtained. Preliminary life testing has been carried out at room temperature on a few laser diodes (λ = 1.5 μm). Operation at constant current for several thousand hours has been achieved with no change in the threshold current.

Reprinted with permission from J.Electron.Mater. 1983, $\underline{12}$(2):371, published by the Metallurgical Society of AIME, Warrendale, PA, USA.

Reif, R. and Dutton, R.W

COMPUTER SIMULATION IN SILICON EPITAXY

    A computer model capable of simulating epitaxial doping
profiles resulting from a variety of deposition conditions,
including time-varying dopant gas flows during growth, is
described. The model takes into account gas-phase dynamics,
physicochemical processes at the growing epitaxial surface, and
thermal redistribution of impurities during epitaxial growth. The
techniques employed to obtain experimentally the values of the
parameters in the computer model are described. The numerical
implementation of the model in the process simulation program SUPREM
is also described. Finally, numerical; simulations using the model
are compared to epitaxial doping profiles obtained experimentally.

J.Electrochem.Soc. 1981, $\underline{128}$(4):909. Reprinted by permission of the
publisher, The Electrochemical Society Inc., Pennington, NJ, USA.

Robbins, D.J., Pidduck, A.J., Hardeman, R.W., Gasson, D.B.,
Pickering, C., Daw, A.C., Chew, N.G., Cullis, A.G., Johnson, M. and
Jones, R.

    This paper reports the first detailed study of Si epitaxy by *in
situ* measurement of diffuse laser light scattering (LLS) from the
growing surface. 488 nm radiation from an Ar^+ laser was directed on
to the surface of both (001) and (111) Si wafers in the deposition
chamber of a Si MBE system. The diffusely-scattered intensity was
measured during deposition in a backscattering geometry. The
technique is shown to be sensitive to surface changes on the
nanometer scale during substrate cleaning and layer nucleation, and
therefore valuable in characterising epitaxial growth modes.

Reprinted with permission from J.Cryst.Growth 1986 (in press).

Roberts, J.S., Dawson, P. and Scott, G.B.

HOMOEPITAXIAL MOLECULAR BEAM GROWTH OF InP ON THERMALLY CLEANED
{110} ORIENTED SUBSTRATES

    {100} oriented homoepitaxial films of InP have been deposited
by molecular beam epitaxy using the dimer species P_2. The starting
surface for epitaxial growth was established by thermally cleaning
the substrate in a P_2 flux in contrast to the generally adopted
technique of argon-ion sputtering followed by annealing. Both
undoped and Sn-doped InP layers deposited within the temperature
range 770-700 K exhibited 77 K mobilities consistent with a
compensation ratio ~2. The 77 K external photoluminescence yield of

0.6 μm-thick layers was comparable with that from 2 μm-thick molecular beam epitaxial films of the same free-electron concentrations but grown on surfaces cleaned by argon-ion sputtering.

Reprinted with permission from Appl.Phys.Lett. 1981, 38(11):905, published by the American Institute of Physics, New York.

Robertson, D.S.

A STUDY OF THE FLOW PATTERNS IN LIQUIDS USING A MODEL CZOCHRALSKI CRYSTAL GROWING SYSTEM

The results of an investigation of the various modes of flow which could be produced in the liquid during Czochralski crystal growth are presented. These indicate that effective liquid movement produced by crystal rotation is dependent upon crystal diameter and upon temperature. Such movement appears to be inhibited by any marked thermal diffusion of the liquid.

Reprinted with permission from Brit.J.Appl.Phys. 1966, 17:1047, published by Institute of Physics, Bristol, UK.

Roth, W., Krautle, H., Krings, A. and Beneking, H.

LASER STIMULATED GROWTH OF EPITAXIAL GaAs

Stimulated growth of single crystalline GaAs has been obtained by irradiation of (100) oriented GaAs substrates inside an MOCVD reactor with a pulsed Nd-YAG laser.

Process temperatures have been varied between 540°C and 360°C. In the non-irradiated areas, below 480°C substrate temperature the growth rate decreases rapidly, whereas in the irradiated part of the substrate epitaxial layers could be grown in the whole temperature range investigated. Below 450°C, the growth is reaction limited.

Reprinted with permission from Materials Research Society Symp.Proc. 1983, 17:193, Copyright 1983 Elsevier Science Publishing Co.Inc.

Rozgonyi, G.A., Deysher, R.P. and Pearce, C.W.

THE IDENTIFICATION, ANNIHILATION AND SUPPRESSION OF NUCLEATION SITES RESPONSIBLE FOR SILICON EPITAXIAL STACKING FAULTS

A study of the formation of epitaxial stacking faults in 2 in. diameter, dislocation-free (111) silicon wafers used in the fabrication of standard buried collector transistors has been made.

The nucleation sites for the epitaxial faults are introduced during
the initial oxidation of the wafer and are correlated with the
presence of a high density of shallow, flat-bottomed, saucer-shaped
etch pits. The saucer pits are selectively annihilated in the
diffused or implanted regions during the fabrication of the Sb-doped
buried collectors. For the ion-implanted process the annihilation
of saucer pits extends laterally from 50 to 100 μm beyond the
boundaries of the collectors. Following epitaxial growth, epitaxial
stacking faults, at a density of 10^4 cm^{-2}, are only found in those
nonburied layer regions which have a saucer pit density of 10^6-10^7
cm^{-2} before epitaxy. Therefore, epi stacking faults are not found
in ion-implanted material with a separation between buried
collectors of 100 μm or less. Most 3 in. diameter, and the central
regions of 2 in. diameter wafers do not have saucer pits or epi
stacking faults. This is attributed to an IiIn situ gettering of
nucleation sites by SiO_2 precipitates, which are known to form in
wafers, or regions of wafers, with a high initial oxygen
concentration. Additional procedures for deliberately suppressing
or gettering the nucleation sites are presented. These include
deliberate abrasion, deposition of strained Si_3N_4 layers,
introduction of misfit dislocations, and the use of an Sb ion
implant, which are performed on the back side of wafers before the
initial masking oxidation.

J.Electrochem.Soc. 1976, 123:1910. Reprinted by permission of the
publishers, The Electrochemical Society Inc., Pennington, NJ, USA.

Russell, G.J., Waite, P. and Woods, J.

ELECTRICALLY ACTIVE GRAIN BOUNDARIES IN POLYCRYSTALLINE ZINC SELENIDE

An investigation is reported of the electrical properties of
grain boundaries in polycrystalline ZnSe grown by chemical vapour
deposition from Zn and H_2Se. The material hyas been studied in the
SEM using the EBIC and CL modes. The EBIC contrast observed is
attributed to band bending at the grain boundaries, which is
probably associated with impurity segregation. The
cathodoluminescence of the bulk grains is characterised by an edge
emission band at 4650A and the copper-red band at 6300A.
Examination in the CL mode shows that both the edge and red emission
are suppressed at the grain boundaries. This may be attributable to
the segregation of a hydride species (known to be present) or
transition metal impurity at the boundaries.

Reprinted with permission from Proceedings of the 2nd Microscopy of
Semiconducting Materials Conference, Oxford 1981, published by
Institute of Physics, Bristol, UK.

Schaake, H.F., Tregilgas, J.H., Lewis, A.J. and Everett, P.M.

LATTICE DEFECTS IN (HgCd)Te: INVESTIGATIONS OF THEIR NATURE AND
EVOLUTION

Brief summaries of the precipitation of tellurium in (HgCd)Te
during the quench from the recrystallisation anneal of the solid
state recrystallisation process and the subsequent dissolution of
these precipitates during the postanneal are given. It is shown
that if vacancies and interstitials are considered to be the only
important point defects in the lattice, the observed mechanisms of
precipitation and annihilation require that tellurium in excess of
stoichiometry must be accommodated by metal vacancies. Diffusion
from the mercury ambient during postanneal then must occur by metal
interstitials. Enhanced self-diffusion on the metal sublattice is
shown to occur along grain and subgrain boundaries. The most
significant source of dark current in CID imagers fabricated on
(Hg,Cd)Te is shown to be a sufficiently high dislocation density in
a pixel.

Reprinted with permission from J.Vac.Sci.Technol 1983, $\underline{A1}$(3):1625,
published by the American Institute of Physics, New York, © 1983,
American Vacuum Society.

Schaake, H.F. and Tregilgas

PRECIPITATION OF TELLURIUM IN (Hg,Cd)Te ALLOYS

The precipitation of tellurium in alloys of $Hg_{1-x}Cd_xTe$ with x =
0.2 to 0.3 prepared under conditions of excess tellurium by the
solid state recrystallisation method have been studied using
chemical defect etching and transmission electron microscopy.
Precipitation of tellurium occurred during the quench from the
recrystallisation anneal. During the initial part of the quench,
precipitation occurred as a result of nucleation of precipitates on
pre-existing dislocations. This precipitation resulted in the
formation of dislocation helices with precipitates in the interior
of the helix loops. Both monoclinic and trigonal phases of
tellurium were found in these precipitate defects, as well as the
twinned matrix phase. Later in the quench, an additional nucleation
mechanism, either homogeneous or hetergeneous on an impurity occurs
resulting in the formation of complex defects in the bulk. During a
subsequent post-anneal at temperatures of approximately 300°C, the
tellurium precipitates dissolve, and interstitial perfect and
faulted dislocation loops are formed. Both the precipitation
process and the subsequent post-annealing result in considerable
multiplication of dislocations.

Reprinted with permission from J.Electron.Mater. 1983, $\underline{12}$(6):931, published by the Metallurgical Society of AIME, Warrendale, PA, USA.

Schaake, H.F., Trgilgas, J.H., Beck, J.D., Kinch, M.A. and Gnade, B.E.

THE EFFECT OF LOW TEMPERATURE ANNEALING ON DEFECTS, IMPURITIES AND ELECTRICAL PROPERTIES OF (Hg,Cd)Te

Many methods for the preparation of (Hg,Cd)Te alloys rely on a low temperature processing step to convert the as-grown p-type material to p-type, or to otherwise adjust the concentration of native acceptors. During this anneal, tellurium precipitates in the material are annihilated by in-diffusing mercury, resulting in a substantial multiplication of dislocations. For substantially long anneals (> 1 day at 270°C) the depth of the p-n junction is found to vary as the square root of the anneal time and inversely as the square root of the excess tellurium concentration. Rapidly diffusing impurities such as silver are gettered out of the skin and into the remaining vacancy-rich core. The kinetics of these processes are analysed for self-diffusion on the metal sublattice involving only vacancies, only interstitials and for a mixed vacancy-interstitial model. Comparison with experimental data shows best agreement with the mixed interstitial-vacancy model.

Reprinted with permission from J.Vac.Sci.Technol. 1985, A3(1):143, published by the American Institute of Physics, © 1985, American Vacuum Society.

Scheel, H.J. and Schulz-Dubois, E.O.

FLUX GROWTH OF LARGE CRYSTALS BY ACCELERATED CRUCIBLE-ROTATION TECHNIQUE

A novel stirring technique is described which, in conjunction with localised cooling, permits control of nucleation in closed crucibles for crystal growth from high-temperature solutions.

Reprinted with permission from J.Cryst.Growth 1971, 8:304, published by North Holland Publishing Co., Amsterdam.

Schmidt, P.F. and Pearce, C.W.

A NEUTRON ACTIVATION ANALYSIS STUDY OF THE SOURCES OF TRANSITION GROUP METAL CONTAMINATION IN THE SILICON DEVICE MANUFACTURING PROCESS

A survey is given of the sources, species and quantities of transition group metals residing on starting silicon substrates and

introduced during the device manufacturing process. The survey
begins with the polysilicon feed for silicon ingot fabrication and
extends through all customary device fabrication steps, such as
wafering, polishing, oxidations, epitaxy, ion implantation, etc.
High temperature furnace operations in the absence of HCl, such as
steam oxidations, are presently the major source of contamination if
all other steps of the fabrication process are under full control.
However, contaminated polishing slurries, contaminated handling
tools or carriers, impure SiC boats or susceptors, or choice of
unsuitable materials for beam catcher plates, etc. in ion
implantation, can also lead to trouble. several options are
available to avoid the contamination introduced during high
temperature treatments, and these are reviewed.

J.Electrochem.Soc. 1981, <u>128</u>(3):630. Reprinted by permission of the
publishers, The Electrochemical Society Inc., Pennington, NJ, USA.

Seager, C.H.

THE ELECTRONIC PROPERTIES OF SEMICONDUCTOR GRAIN BOUNDARIES

Grain boundaries play an important role in determining the
operating characteristics of devices such as varistors and thin film
solar cells and transistors. In the last several years detailed
one-electron expressions have been developed for the transport
coefficients of majority and minority carriers at grain boundaries.
These theories have been successful in predicting the measured
electrical properties of bicrystals in lightly doped silicon. The
status of our understanding in this area will be reviewed in some
detail. Recent extensions of these calculations which have been
necessary to explain measurements on grain boundaries in
degenerately doped GaAs will also be reviewed.

Reprinted by permission of the publisher from "The electronic
properties of semiconductor grain boundaries", p185, in "Grain
Boundaries in Semiconductors", Eds. C.H.Seager and P.Lemy, Copyright
1982 by Elsevier Science Publishing Co.Inc.

Series, R.W. and Barraclough, K.G.

CARBON CONTAMINATION DURING GROWTH OF CZOCHRALSKI SILICON

Detailed measurements of the axial distribution of carbon
within eight 2 and 3 inch diameter silicon crystals grown in a low
pressure Czochralski puller have been made. Comparison of the data
with a modified version of the normal freeze equation permits
evaluation of the rate of gain or loss of carbon from the melt
during growth. Contamination rates during growth of about 0.1 μg

s^{-1} were measured and in the absence of other sources of contamination, this would limit the carbon content of the crystals to about $2x10^{15}$ atoms cm^{-3} at the seed and $2x10^{16}$ atoms cm^{-3} at a fraction solid of 0.7. The measured carbon contents of the crystals were generally higher than this value indicating that contamination of the crucible, starting charge and melt prior to growth were the dominant source of carbon. No evidence was found to suggest a loss of carbon from the melt through the formation of carbon monoxide.

Reprinted with permission from J.Cryst.Growth 1982, 60:212, published by North Holland Publishing Co., Amsterdam.

Series, R.W. and Barraclough, K.G.

CONTROL OF CARBON IN CZOCHRALSKI SILICON CRYSTALS

Measurements of the rate of carbon contamination of a silicon melt in a low pressure Czochralski puller have been made by growing crystals as a series of repeated dips. The effect of a gas baffle placed between the heater and the melt has also been examined. It was found that for the levels of carbon generally encountered, contamination from the gas ambient is minor compared to the contamination of the charge which occurs before growth starts. The presence of a gas baffle placed between the heater and the melt was found to increase the rate of contamination of the melt.

Reprinted with permission from J.Cryst.Growth 1983, 63:219, published by North Holland Publishing Co., Amsterdam.

Shaw, D.W.

ADVANCED MULTILAYER EPITAXIAL STRUCTURES

Epitaxial growth is being applied to form increasingly complex multilayer/multimaterial structures that have the potential for major advances in electronic and optical devices. These structures, which include single crystal multimaterial composites, superlattices and metastable compositions or phases, are discussed with emphasis on their method of preparation and potential applications.

Reprinted with permission from J.Cryst.Growth 1983, 65:444, published by North Holland Publishing Co., Amsterdam.

Shimada, T., Terashima, K., Nakajima, H. and Fukuda, T.

GROWTH OF LOW AND HOMOGENEOUS DISLOCATION DENSITY GaAs CRYSTAL BY IMPROVED LEC TECHNIQUE

An improved LEC technique of enhanced heating of B_2O_3 encapsulant layer through the windows of susceptor cylinder is described. By this technique, low temperature gradients of 10 – 25°C/cm in the axial direction and 5°C/cm in the radial direction were achieved. Two-inch diameter crystals grown by this technique with a slow cooling rate exhibited quite low dislocation density of 8×10^3 to low 10^4 cm^{-2} order from front to tail of the crystal.

Reprinted with permission from Jpn.J.Appl.Phys. 1984, 23(1):L23,

Sinharoy, S., Hoffman, R.A., Rieger, J.H., Takei, W.J. and Farrow, R.F.C.

EPITAXIAL GROWTH OF LANTHANIDE TRIFLUORIDES BY MBE

The first epitaxial growth of a hexagonal structure lanthanide trifluoride on a semiconductor is reported: LaF_3 on Si(111). LEED, RHEED, and X-ray diffraction studies show that the c-axis of LaF_3 is normal to the Si(111) plane. A similar epitaxial relation is expected for other members of the lanthanide trifluoride family: CeF_3, PrF_3 and NdF_3 which are isomorphous with LaF_3 and have a closer (basal plane) lattice parameter match to Si(111). The hard, water insoluble nature of these materials makes them more practicable than group II fluorides as exploratory films for semiconductor passivation, gate insulator and epitaxial interlayers.

Reprinted with permission from J.Vac.Sci.Technol. 1985, B3(2):722, published by the American Institute of Physics, New York, © 1985, American Vacuum Society.

Siskos, S., Fontaine, C. and Munoz-Yague, A.

GaAs/$(Ca,Sr)F_2$/(001)GaAs LATTICE-MATCHED STRUCTURES GROWN BY MOLECULAR BEAM EPITAXY

Epitaxial growth of GaAs/$(Ca,Sr)F_2$/GaAs structures by means of molecular beam epitaxy has been demonstrated. It has been shown that it was possible to grow layers with good crystalline quality and no noticeable interdiffusion at the interfaces. In addition, the uppermost GaAs layers present interesting electrical properties, exhibiting electron Hall mobilities of about 1300cm^2/Vs.

Reprinted with permission from Appl.Phys.Lett. 1984, 44:146, published by the American Institute of Physics, New York.

Skolnick, M.S., Brozel, M.R., Reed, L.J., Grant, I., Stirland, D.J. and Ware, R.M.

INHOMOGENEITY OF THE DEEP CENTER EL2 IN GaAs OBSERVED BY DIRECT INFRA-RED IMAGING

The distribution of the dominant deep trap EL2 in 7.5 cm diameter crystals of semi-insulating GaAs is studied by whole slice infrared imaging. Very significant fluctuations in the neutral EL2 concentration ([EL2]°) are observed. The different sorts of fine structure, namely cell structure and bands of high infrared absorption ("sheets" and "streamers") lying in (110) planes running down the <001> growth directions, are described.

Reprinted with permission from J.Electron.Mater. 1984, 13(1):107, published by the Metallurgical Society of AIME, Warrendale, PA, USA.

Stirland, D.J., Augustus, P.D. and Straughan, B.W.

THE IDENTIFICATION OF SAUCER-PIT (S-PIT) DEFECTS IN GaAs

Samples of gallium arsenide from liquid encapsulated Czochralski grown ingots, doped with either tellurium or selenium to carrier concentrations ~ 10^{18} cm^{-3} revealed shallow pits (S-pits) by etching. Although the S-pits were randomly distributed throughout the matrix, areas of high densities were associated with dislocations. This observation was utilised to identify the types of defects which became S-pits when etched. Transmission electron microscope specimens of etched material were examined in the dislocation regions, and showed directly that faulted loops with Frank partials b = $1/3a_0$<111> containing precipitate particles could become S-pits. It was further deduced from the combined optical and electron microscope observations that both faulted {111} and unfaulted {110} dislocation loops became S-pits provided they contained precipitate particles.

Reprinted with permission from J.Mater.Sci. 1978, 13:657, published by Chapman and Hall, London.

Sugawara, K.

SILICON EPITAXIAL GROWTH BY ROTATING DISK METHOD

By a rotating disk method which had quantitatively been treated in fluid motion, silicon epitaxial deposition using the $SiCl_4$-H_2 system was effected at various susceptor rotation speeds and various pressures. Calculation of the growth rate was initially made with a one-dimensional model of an infinite diameter disk without

considering natural convection caused by local heating. This was
followed by a numerical solution of Navier-Stokes equations,
continuity equation and equation of energy of cylindrical
coordinates, using a three-dimensional model and taking natural
convection into consideration. With the three-dimensional model
including natural convection, a quantitative description can be
given of the influence of susceptor rotation speed and pressure on
the growth rate at a temperature higher than 1150°C.

J.Electrochem.Soc. 1972, 119:1749. Reprinted by permission of the
publishers, The Electrochemical Society Inc., Pennington, NJ, USA.

Sullivan, P.W., Farrow, R.F.C. and Jones, G.R.

INSULATING EPITAXIAL FILMS OF BaF_2, CaF_2 AND $Ba_xCa_{1-x}F_2$
GROWN BY MBE ON InP SUBSTRATES

Thin films ($\leqslant$5000 A) of BaF_2, CaF_2 and $Ba_xCa_{1-x}F_2$ have been
grown onto InP (001) substrates in a vacuum locked MBE system.
Electron diffraction was used to monitor film nucleation and growth
at a variety of substrate temperatures. Subsequent ex-situ analysis
included X-ray diffraction, electron microscopy and Auger sputter
profiling. In addition, capacitance-voltage and current-voltage
analyses were performed on MIS sandwich devices formed by
evaporating aluminium onto the semiconductor-fluoride samples.
Deposition of fluorides onto room temperature substrates resulted in
the growth of smooth, pinhole free, stoichiometric polycrystalline
films with little or no preferred orientation. TEM analysis
indicates a grain size of the same order as the film thickness (~
1000 A). Deposition of BaF_2 and CaF_2 onto cleaned, well-ordered
(001) InP held at temperatures above 200°C resulted in
single-crystal, heteroepitaxial growth. In the case of CaF_2, the
increase in lattice mismatch on cooling to room temperature resulted
in crazing of the epitaxial layer. Film resistivity values around
10^{12}-10^{13} Ω cm and breakdown strengths of $5x10^5$ V cm^{-1} have been
achieved for both polycrystalline and single-crystal layers. In
initial experiments on the growth of $Ba_xCa_{1-x}F_2$ alloys onto (001)
InP, epitaxial films of a single-phase cubic alloy with x ~ 0.2 were
obtained.

Reprinted with permission from J.Cryst.Growth 1982, 60:403,
published by North Holland Publishing Co., Amsterdam.

Sullivan, P.W.

GROWTH OF SINGLE CRYSTAL SrF_2(001)/GaAs(001) STRUCTURES BY
MOLECULAR BEAM EPITAXY

Epitaxial SrF_2 layers were grown onto GaAs(001) at $T > 250°C$ by molecular beam epitaxy. The dominant growth mechanism appears to be by three-dimensional island growth although significant ordering of the surface occurs on annealing the layers at $T > 400°C$. Growth of epitaxial GaAs (001) at 540°C by molecular beam epitaxy onto the SrF_2 films has also been demonstrated. The results suggest that stacked, single crystal, alternating layers of GaAs (001) and SrF_2 (001) should be amenable to growth by molecular beam epitaxy.

Reprinted with permission from Appl.Phys.Lett. 1984, 44:140, published by the American Institute of Physics, New York.

Sullivan, P.W., Bower, J.E. and Metze, G.M.

GROWTH OF SEMICONDUCTOR/INSULATOR STRUCTURES - GaAs/FLUORIDE/GaAs(001)

Thin single-crystal layers (~ 100-2000 A) of (001) oriented SrF_2 and CaF_2 have been grown onto clean MBE-grown GaAs (001) surfaces. RED patterns suggest a significant difference in the interaction of CaF_2 and SrF_2 with clean reconstructed GaAs (001). Three-dimensional behavior is suggested in the initial nucleation stages of CaF_2 growth, although significant ordering occurs for film thickness > 100 A. Nucleation of SrF_2 on $I(2x8)$ reconstructed GaAs surfaces results in a well-ordered reconstructed (4x2) surface up to film thicknesses of ~ 100 A, after which a disordered (1x1) unreconstructed surface is apparent. GaAs has been grown onto MBE prepared fluoride surfaces. At sufficiently low growth rates on CaF_2, RED patterns indicate well ordered surfaces with (4x4) reconstruction. Increased deposition rates lead to twinned GaAs growth. Photoluminescence studies of undoped GaAs layers on SrF_2 reveal a relatively sharp exciton structure with a full width at half-maximum (FWHM) of 4 meV. Auger depth profile experiments suggest little or no interdiffusion between the GaAs and SrF_2 layers while the presence of both Ga and As is detected in the CaF_2.

Reprinted with permission from J.Vac.Sci.Technol. 1985, B(3):500, published by the American Institute of Physics, New York, © 1985 American Vacuum Society.

Taniguchi, M. and Ikoma, T.

VARIATION OF THE MIDGAP ELECTRON TRAPS (EL2) IN LIQUID ENCAPSULATED CZOCHRALSKI GaAs

We have characterized the midgap electron trap (so far believed to be EL2) in liquid encapsulated Czochralski GaAs by measuring deep level transient spectroscopy spectra, and found more than two midgap electron traps which can be classified into two groups. A trap belonging to the first group is rather stable in its properties and the other is unstable. The capture cross section of the levels in the latter group varies continuously in depth from the surface and also is changed by thermal annealing. We discuss these characteristics in connection with growth conditions.

Reprinted with permission from J.Appl.Phys. 1983, 54:6448, published by the American Institute of Physics, New York.

Tiller, W.A.

SOME UNRESOLVED THEORETICAL PROBLEMS IN CRYSTAL GROWTH

Attention is focused on three areas: (1) the need to begin using potential energy functions that include the many-body effect, rather than just the two-body effect, if one wishes to meaningfully assess the energetics and structural features of surfaces and interfaces. (2) The importance of critical surface adsorption on surface mobility and the need to control this over a wide range of temperature, pressure and deposition conditions to assure high quality crystal growth. (3) Some consequences of adsorption and interface fields on interface instability.

Reprinted with permission from J.Cryst.Growth 1984, 70:13, published by North Holland Publishing Co., Amsterdam.

Tom, H.W.K., Mate, C.M., Zhu, X.D., Crowell, J.E., Heinz, Somorjai, G.A., and Shen, Y.R.

SURFACE STUDIES BY OPTICAL SECOND-HARMONIC GENERATION: THE ADSORPTION OF O_2, CO_2 AND SODIUM ON THE Rh(111) SURFACE

Optical second-harmonic generation is used to study the adsorption of atomic and diatomic species on a well characterised Rh(111) crystal surface in ultra-high vacuum. The results correlate well with those obtained by other surface probes.

Reprinted with permission from Phys.Rev.Lett. 1984, 52(5):348, published by the American Physical Society, New York.
Tregilgas, J., Beck, J. and Gnade, B.

TYPE CONVERSION OF (Hg,Cd)Te INDUCED BY THE REDISTRIBUTION OF RESIDUAL ACCEPTOR IMPURITIES

Bulk (Hg,Cd)Te grown by the quench-anneal process has been found to contain residual acceptor impurities which are normally depleted from the n-type surface by an internal gettering mechanism during low temperature postannealing in saturated Hg vapor. When the n-type slices are postannealed for much longer periods, the Te saturated core is annihilated and the gettered residual acceptor impurities can redistribute into the surface region. If upon homogenisation the residual acceptor concentration exceeds the residual donor concentration, the material will convert to p-type. A new method for applying neutron activation analysis (NAA) to (Hg,Cd)Te has been used to identify this acceptor impurity as Cu at levels below 1 ppm. An extension of the NAA technique has employed quartz encapsulated silicon samples to demonstrate Cu may come from either the quartz or annealing environment.

Reprinted with permission from J.Vac.Sci.Technol. 1985, A3(1):150, published by the American Institute of Physics, © 1985 American Vacuum Society.

Tsang, W.T., Miller, R.C., Capasso, F. and Bonner, W.A.

HIGH QUALITY InP GROWN BY MOLECULAR BEAM EPITAXY

We report the first high quality InP grown by molecular beam epitaxy (MBE). The undoped InP layers are InI type with residual impurity concentrations $\sim 5\times10^{14}$ - $\sim 5\times10^{15}$ cm^{-3}. Fine structure attributed to polariton, neutral donor-exciton (D°-x), neutral donor-hole (D°-h), neutral acceptor-exciton (A°-x) transitions at the exciton edge and neutral donor-neutral acceptor (D°-A°) transitions are clearly resolved in the low-temperature (5K) photoluminescence spectra with a linewidth of <1 meV for D°-x as has been observed with high-purity InP layers grown by other methods. Elemental In and P (red phosphorus) were used as the primary molecular beam sources. The growth temperature has a very significant effect on the quality of the InP layers. The advanced design of the present MBE system employed for growing III-V compound semiconductors containing P from elemental red phosphorus is also described.

Reprinted with permission from Appl.Phys.Lett. 1982, 41(5):467, published by the American Institute of Physics, New York.

Tsang, W.T.

GROWTH OF InP, GaAs and $InS_{0.53}GaS_{0.47}As$ BY CHEMICAL BEAM EPITAXY

A new epitaxial growth technique, chemical beam epitaxy (CBE), was demonstrated and investigated with the growth of InP and $InS_{0.53}GaS_{0.47}As$ GaAs. In this technique, all the sources were gaseous group III and group V alkyls. The In and Ga were derived by the pyrolysis at the heated substrate surface of either trimethyl-indium (TMIn) or triethylindium (TEIn), and trimethylgallium (TMGa) or triethylgallium (TEGa), respectively. The As_2 and P_2 were obtained by thermal decomposition of triethylphosphine (TEP) and trimethylarsine (TMAs) in contact with heated Ta or Mo at 950-1200°C, respectively. Unlike conventional vapor phase epitaxy, in which the chemicals reached the substrate surface by diffusing through a stagnant carrier gas boundary layer above the substrate, the chemicals in CBE were admitted into a high vacuum growth chamber and impinged directly line of sight onto the heated substrate surface in the form of molecular beams. The beam nature of CBE resulted in very efficient use of the impinging chemicals and allowed the utilisation of mechanical shutters. A similar gas handling system to that employed in conventional metalorganic chemical vapor deposition (MOCVD) with precision electronic mass flow controllers was used for controlling the flow rates of the various gases admitted into the growth chamber. Growth rates as high as 3-5 mm/h have been achieved for both InP and GaAs. The InP and GaAs epilayers grown have smooth surfaces and comparable optical quality to bulk substrates. Since TMIn and TMGa emerged as a single mixed beam, spatial composition uniformity was automatically achieved without the need of substrate rotation in the $InS_{0.53}GaS_{0.47}As$ epilayers grown. Lattice mismatch $\Delta a/a \leqslant 4 \times 10^{-4}$ was obtained.

Reprinted with permission from J.Vac.Sci.Technol. 1985, B3(2):666, published by the American Institute of Physics, © 1985, American Vacuum Society.

Tu, C.W., Forrest, S.R. and Johnston Jr., W.D.

EPITAXIAL InP/FLUORIDE/InP(001) DOUBLE HETEROSTRUCTURES GROWN BY MOLECULAR BEAM EPITAXY

We report the first epitaxial semiconductor-dielectric-semiconductor (SDS) double heterostructures using the III-V compound semiconductor InP. The samples, $InP/CaF_2/InP(001)$ and $InP/Ba_xSr_{1-x}F_2/InP(001)$, were grown by molecular beam epitaxy and have lattice mismatches of - 6.9% and + 2.0% respectively. *In situ* high-energy electron diffraction showed that the initial stage of epitaxy of the InP/fluoride structure, unlike that of the fluoride/InP structure, exhibits pseudomorphism. Analysis of the electrical properties of SDS devices with an insulator thickness of ~ 100 A indicates both Ohmic and trap-assisted tunneling conduction.

Reprinted with permission from Appl.Phys.Lett. 1983, 43(6):569, published by the Institute of Physics, New York.

Tu, C.W., Sheng, T.T., Read, M.H., Schlier, A.R., Johnson, J.G., Johnston Jr., W. D. and Bonner, W.A.

GROWTH OF SINGLE-CRYSTALLINE EPITAXIAL GROUP II FLUORIDE FILMS ON InP(001) BY MOLECULAR-BEAM EPITAXY

Twin-free, single-crystalline, lattice-mismatched as well as lattice-matched, epitaxial dielectric films of group II cubic fluorides (SrF_2, CaF_2 and $Ba_xSr_{1-x}F_2$) have been grown on InP(001) substrates by molecular-beam epitaxy. The InP(001) surface was cleaned in vacuum by heating uner phosphorus overpressure until a well-ordered and stoichiometric surface was obtained. The film growth of MF_2, where M is Ba, Sr, Ca or Ba_xSr_{1-x} was followed by reflection high energy electron diffraction (RHEED). The diffraction patterns indicate a parallel epitaxial relationship: MF_2(001);InP(001) and MF_2[110];InP[110]. At low growth temperature ($\sim 250°C$) and low growth rate we could obtain twin-free single-crystalline CaF_2 films. Transmission electron microscopy show the absence of any grains. At higher growth temperature ($\sim 350°C$) we could obtain twin-free single-crystalline SrF_2, CaF_2 and $Ba_xSr_{1-x}F_2$ films. The latter can be lattice-matched to InP(001) at room temperature.

J.Electrochem.Soc. 1983, 130:2081. Reprinted by permission of the publishers, The Electrochemical Society Inc., Pennington, NJ, USA.

Tung, R.T., Bean, J.C., Gibson, J.M., Poate, J.M. and Jacobson, D.C.

GROWTH OF SINGLE-CRYSTAL $CoSi_2$ ON Si(111)

Single-crystal $CoSi_2$ films have been grown under ultrahigh vacuum conditions on Si (111) by both standard deposition and molecular beam epitaxy techniques. Films were analysed by Rutherford backscattering spectroscopy and channeling, transmission electron microscopy and low energy electron diffraction. The films are free of grain boundaries but are rotated 180° about the normal to the Si surface. The crystalline perfection, as measured by channeling, is the best yet reported for an epitaxial silicide system. The expected hexagonal misfit dislocation arrays, along with a coarser triangular defect structure, are confined to the plane of the interface.

Reprinted with permission from Appl.Phys.Lett. 1982, 40(8):684, published by the American Institute of Physics, New York.

Tung, R.T., Poate, J.M., Bean, J.C., Gibson, J.M. and Jacobson, D.C.

EPITAXIAL SILICIDES

The formation and structures of epitaxial $CoSi_2$, $NiSi_2$, Pd_2Si and PtSi films on silicon are reviewed. Polycrystalline films of reasonable epitaxial quality can be grown with sharp interfaces on Si(111) by conventional deposition and heating techniques. The interfaces on (100) substrates, however, are faceted. Perfect single-crystal $CoSi_2$ films can be grown on Si(111) substrates by ultrahigh vacuum techniques using conventional deposition and annealing at about 900°C and molecular beam epitaxy codeposition at about 600°C. The single-crystal films are rotated 180°C with respect to the substrate. Novel defect arrays are observed at the interface. Resistivities in epitaxial films are lower than those reported in polycrystalline layers. High quality silicon films can be grown on the silicides to form semiconductor/metal/ semiconductor heterostructures.

Reprinted with permission from Thin Solid Films 1982, 93:77, published by Elsevier Sequoia, SA, Lausanne, Switzerland.

Tunnicliffe, J., Blackmore, G.W., Irvine, S.J.C., Mullin, J.B. and Holland, R.

SIMS ANALYSIS OF IMPURITIES AT CMT/CdTe HETEROSTRUCTURE INTERFACES

Major and impurity element depth profiling has been carried out using SIMS on a heterostructure of CMT/CdTe grown by MOVPE onto a CdTe substrate. Li and Al peaks were only observed at the substrate/layer interface and have been attributed to contamination on the substrate surface.

Reprinted with permission from Mater.Lett. 1984, 2(5a):393, published by North Holland Physics Publishing, Elsevier, Amsterdam.

Vere, A.W., Cole, S. and Williams, D.J.

THE ORIGINS OF TWINNING IN CdTe

The use of cadmium telluride is limited by the occurrence of twins in the material. Mechanical deformation experiments and studies of the growth interface suggest that these arise through growth accidents occurring on the liquid-solid interface rather than by mechanical deformation of the solidified ingot. The incidence of twinning is found to be higher in Czochralski-grown material than that grown by other techniques and there is a consequent reduction in grain-size due to the formation of incoherent boundaries at twin intersections. From comparisons with other II-VI and III-V compounds it is concluded that the twinning is an intrinsic feature of the CdTe chemical bond structure and can only be totally eliminated by changing the ionicity of the material. Nevertheless, the twin density can be reduced by the use of low growth rates and the avoidance of interface perturbations.

Reprinted with permission from J.Electron.Mater. 1983, 12(3):551, published by the Metallurgical Society of AIME, Warrendale, PA, USA.

Vere, A.W., Steward, V., Jones, C.A., Williams, D.J. and Shaw, N.

GROWTH OF CdTe BY SOLVENT EVAPORATION

     Recent improvements to the Solvent Evaporation (SE) process are described. It is shown that good crystallinity is achievable provided that the solid-liquid interface is convex with respect to the solid and that the pyrolysed carbon coating of the ampoule adheres firmly to the wall. Breakdown of the coating leads to carbon-rich inclusions in the material and to dissolved carbon in the crystal lattice. C-V measurements indicate that carbon-rich areas are higher n-type than carbon-free regions, suggesting that carbon is a residual donor impurity in CdTe.

Reprinted with permission from J.Cryst.Growth 1985, 72:97, published by North Holland Publishing Co., Amsterdam.

Veuhoff, E., Pletschen, W., Balk, P. and Luth, H.

METALORGANIC CVD OF GaAs IN A MOLECULAR BEAM SYSTEM

     The feasibility of growing GaAs using molecular beams of $Ga(CH_3)$ and AsH_3 in an UHV system is shown. Using cold sources placed outside the system these gaseous components are introduced into the growth apparatus and directed towards the substrate surface using capillary tubes with diffuser chambers. Films obtained in this process are of uniform thickness; their growth rate is limited by the supply of the Ga compound. Without purposely injecting dopant impurities the layers are p-type and compensated.

Reprinted with permission from J.Cryst.Growth 1981, 55:30, published by North Holland Publishing Co., Amsterdam.

Wahl, G.

HYDRODYNAMIC DESCRIPTION OF CVD PROCESSES

     A CVD process with a two-component gas mixture A-B where only the component B is deposited is solved numerically for an axisymmetric stagnation flow geometry assuming the mass fraction of

B on the deposition surface to have a fixed value Y_{Bd}. A comparison of the calculated flow lines with visualisation experiments shows satisfactory agreement. The agreement between calculations and SiO_2 and Si_3N_4 deposition experiments ($SiH_4 + O_2 \rightarrow SiO_2$, $SiH_4 + NH_3 \rightarrow Si_3N_4$) was good for some deposition ranges. In these ranges the deposition profiles could be calculated with only one fitting parameter (Y_{Bd}).

Reprinted with permission from Thin Solid Films 1977, 40:13, published by Elseiver Sequoia, Lausanne, Switzerland.

Wahl, G. Hoffmann, R.

CHEMICAL ENGINEERING WITH CVD

Hydrodynamic problems in small and large CVD-reactors are discussed. They encompass the compensation of the gas atmosphere in CVD-reactors, impurity flows and optimal gas compositions as used for coating components. The following methods of investigating these problems are described:

1) Calculations of gas flow and deposition
2) Fume experiments

These investigations were supported by deposition measurements employing coatings of tungsten ($WF_6 + 3H_2 \rightarrow W + 6HF$) and Si_3N_4 ($4NH_3 + 3SiH_4 \rightarrow Si_3N_4 + ...$) coatings.

Reprinted with permission from Rev.Int.Hautes Temper.Refract. 1980, 17:7, published by Masson, Editeur SA, Paris.

Wang, W.I.

MOLECULAR BEAM EPITAXIAL GROWTH AND MATERIAL PROPERTIES OF GaAs AND AlGaAs ON Si(100)

We have grown GaAs and AlGaAs on (100) oriented Si substrates by molecular beam epitaxy. The epitaxial growth was studied in situ by reflection high-energy electron diffraction. Low temperature photoluminescence, Raman scattering and scanning electron microscopy were used to characterise the epitaxial layers. It is shown for the first time that antiphase disorder could be suppressed. The doped AlGaAs grown directly on Si substrates exhibited PL efficiency similar to that of AlGaAs grown on GaAs substrates.

Reprinted with permission from Appl.Phys.Lett. 1984, 44:1149, published by the American Institute of Physics, New York.

Wanklyn, B.M.

THE PRESENT STATUS OF FLUX GROWTH

In this short review, recent work on the following topics is summarised and discussed: (1) complex ions in fluxed melts; (2) crystal growth in aqueous systems related to complexes in solution; (3) habit and number of crystals obtained in relation to the acid/base ratio; (4) control of spontaneous nucleation.

Reprinted with permission from J.Cryst.Growth 1983, $\underline{65}$:533, published by North Holland Publishing Co., Amsterdam.

Warrington, D.H.

FORMAL GEOMETRICAL ASPECTS OF GRAIN BOUNDARY STRUCTURE

The paper sets out the formal definitions of the Bollmann '0 Lattice' and the Coincidence Site Lattice (CSL) derived from it through special or rational rotations and transformations. The operation of displacements of the CSL to produce a Displacement Shift Complete (DSC) lattice defining the boundary structure is then examined. The formalisms derived are applied to a variety of materials structures and orientations.

Reprinted from 'Formal Geometric aspects of grain boundary structure', p1 in 'Grain Boundary Structure and Kinetics', published by ASM, Metals Park, USA.

Warwick, C.A., Brown, G.T., Booker, G.R. and Cockayne, B.

DOPANT INHOMOGENEITY IN CZOCHRALSKI-GROWN INDIUM PHOSPHIDE INGOTS DOPED WITH GERMANIUM

Dopant inhomogeneity in the form of striations in liquid encapsulated Czochralski (LEC) InP ingots doped with Ge in the range $5x10^{16}$ to $1x10^{19}$ cm^{-3} has been studied by scanning electron microscopy (SEM) in the cathodoluminescent (CL) mode and by chemical etching. Striation amplitude, ie the variation in dopant concentration on going from striation to striation, was determined as a function of position in the ingot, mean doping level, growth conditions and facet/non-facet growth. The results give an insight to the thermal environment of the ingot growth and suggest ways of reducing the inhomogeneity.

Reprinted with permission from J.Cryst.Growth 1983, $\underline{64}$:108, published by North Holland Publishing Co., Amsterdam.

Warwick, C.A. and Brown, G.T.

SPATIAL DISTRIBUTION OF 0.68eV EMISSION FROM UNDOPED SEMI-INSULATING GALLIUM ARSENIDE REVEALED BY HIGH RESOLUTION LUMINESCENCE IMAGING

The luminescence properties of undoped semi-insulating liquid encapsulated Czochralski-grown GaAs ingots have been characterised with ~3 μm spatial resolution using a scanning electron microscope based, cathodoluminescence (CL) system. The use of a cryostat operating at ~6 K with cooled PbS and Ge detectors for midgap luminescence imaging has revealed for the first time a depletion in the 0.68 eV emission from zones ~20 μm wide adjacent to individual polygonised dislocation arrays known as cell walls. These midgap luminescence results have been correlated with band edge CL images of cell structures and with dislocation images obtained using double crystal X-ray topography. It is concluded that dislocation arrays attract or getter the defect centers responsible for the 0.68 eV emission in undoped semi-insulating GaAs, as opposed to generating them.

Reprinted with permission from Appl.Phys.Lett. 1985, 46(6):574, published by the American Institute of Physics, New York.

Weber, E.R., Ennen, H., Kaufmann, U., Windscheif, J., Schneider, J. and Wosinski, T.

IDENTIFICATION OF As_{Ga} ANTISITES IN PLASTICALLY DEFORMED GaAs

AsGa antisite defects formed during plastic deformation of GaAs are identified by electron paramagnetic resonance (EPR) measurements. From photo-EPR results it can be concluded that the two levels of this double donor are located near E_c - 0.75 eV and E_v + 0.5 eV. These values are coincident with the Fermi level pinning energies at Schottky barriers. The upper level can be related to the "main electron trap" EL2 in GaAs. Photoluminescence experiments before and after thermal annealing suggest that As_{Ga} defects reduce the near band edge luminescence efficiency. A dislocation climb model is presented which is able to explain As_{Ga} formation during dislocation movement. The production of AsGa antisites during dislocation motion under injection conditions in light emitting devices may thus be connected with degradation of the light output.

Reprinted with permission from J.Appl.Phys. 1982, 53(9):6140, published by the American Institute of Physics, New York.

Westphal, G.H.

CONVECTIVE TRANSPORT IN VAPOR GROWTH SYSTEMS

The interaction of gravity with fluid density gradients greatly influences mass and heat transfer in vapor growth systems. These affects are reviewed by discussing specific examples of closed and open tube chemical or physical vapor transport systems. Both theoretical analysis and experimental results are presented. Emphasis is placed on the closed tube systems since they are often the best characterised and are the best initial candidates for study in a low-g environment (eg space). The open tube systems discussed include halide vapor phase epitaxy of GaAs and the endothermic (cold wall) epitaxial process for Si. The convective stability criteria for open tubes are reviewed and a guide to the mass transfer literature is presented. For terrestrial closed tube systems, two recent results are of particular significance. First, laser Doppler anemometry experiments have demonstrated a complicated detailed flow pattern near the crystal interface. Presumably this effect can strongly influence crystal homogeneity. Second, numerical calculations have shown the well known Klosse and Ullersma transport rate calculations [J Crystal Growth 18 (1973) 167] to be correct only at low Grashof numbers and that such closed tube transport is also Prandtl number dependent.

Reprinted with permission from J.Cryst.Growth 1983, 65:105, published by North Holland Publishing Co., Amsterdam.

Wilkes, J.G.

LOW PRESSURE PROCESSES IN CHEMICAL VAPOUR DEPOSITION OF SILICON OXIDES

The low pressure chemical vapour deposition of silicon oxide from the silane-nitric oxide reaction has been examined under both thermal and plasma excited conditions. Kinetic measurements made over a range of pressures and temperatures have been used to derive values for the activation energy 1.29eV and sorption 0.09eV contributions to this process. The adsorption and desorption steps involved in this multicomponent reaction are related to changes in oxide layer composition as the silane-nitric oxide mol ratio is altered. In applying plasma excitation it is important to use low power to avoid fragmentation of the incoming reactants.

Reprinted with permission from J.Cryst.Growth 1984, 70:271, published by North Holland Publishing Co., Amsterdam.

Williams, G.M., Whitehouse, C.R., Chew, N.G., Blackmore, G.W. and
Cullis, A.G.

AN MBE ROUTE TOWARDS CdTe/InSb SUPERLATTICES

As part of a program aimed at the growth of CdTe/InSb
superlattices, a systematic study has been made of the MBE growth of
heteroepitaxial CdTe layers on (001)InSb using substrate
temperatures, Ts, significantly higher than have been previously
reported. Using a modified two-step growth technique, high quality
layers have been successfully grown at temperatures as high as 310°C
with no evidence of either preferential Cd loss or CdTe/InSb
interdiffusion. The new growth technique is described and
cross-sectional TEM and SIMS data from the grown layers is presented.

Reprinted with permission from J.Vac.Sci.Technol. 1985, B3(2):704,
published by the American Institute of Physics, © 1985, American
Vacuum Society.

Williams, D.J. and Vere, A.W.

TELLURIUM PRECIPITATION IN BULK-GROWN $Cd_xHg_{1-x}Te$

When Te-saturated $Cd_xHg_{1-x}Te$ is cooled from temperatures in
excess of approximately 400°C, precipitation of second-phase Te may
occur. We have studied this precipitation as a function of the
microstructure of the material prior to cooling and as a function of
cooling rate. The results indicate that these parameters control
the balance between heterogeneous and homogeneous nucleation and
that, by careful control of precooling time at high temperature and
coolinng rate, precipitate-free material can be reproducibly
obtained.

Reprinted with permission from J.Vac.Sci.Technol. 1986, A4(4): 2184
published by the American Institute of Physics, © 1986, American
Vacuum Society.

Wright, P.J. and Cockayne, B.

THE ORGANOMETALLIC CHEMICAL VAPOUR DEPOSITION OF ZnS AND ZnSe AT
ATMOSPHERIC PRESSURE

It is shown that thin film single crystal layers of ZnS, ZnSe
and Zn_xSe_{1-x} can be grown on to a variety of substrates by direct
reaction at atmospheric pressure, of dimethyl zinc, hydrogen
sulphide and/or hydrogen selenide, using hydrogen as the carrier
gas. Growth has been observed between 350 and 750°C at growth rates
within the range 0.5 to 10 μm h^{-1} . The layers exhibit a uniformly

high standard of surface morphology and perfection.

Reprinted with permission from J.Cryst.Growth 1982, 59:148,
published by North Holland Publishing Co., Amsterdam.

Wright, P.J., Griffiths, R.J.M. and Cockayne, B.

THE USE OF HETEROCYCLIC COMPOUNDS IN THE ORGANOMETALLIC CHEMICAL
VAPOUR DEPOSITION OF EPITAXIAL ZnS, ZnSe AND ZnO

Heterocyclic compounds are shown to be useful as sources of the
group VI elements in their reactions with dimethylzinc for the
growth of crystalline layers of zinc sulphide, zinc selenide and
zinc oxide in a standard horizontal organometallic chemical vapour
deposition reactor operating at atmospheric pressure. Single
crystal layers of zinc sulphide, zinc selenide and zinc oxide have
been produced on substrates which are closely related in structure
whilst polycrystalline growth of these three compounds has been
demonstrated on a wide variety of substrate materials. The problem
of premature reaction which is observed between group VI hydrides
and dimethylzinc is shown to be insignificant when the heterocyclic
compounds are used in their stead.

Reprinted with permission from J.Cryst.Growth 1984, 66:26, published
by North Holland Publishing Co., Amsterdam.

Yoshida, M., Watanabe, H. and Uesugi, F.

MASS SPECTROMETRIC STUDY OF $Ga(CH_3)_3$ AND $Ga(C_2H_5)_3$ DECOMPOSITION
REACTION IN H_2 AND N_2

Decomposition of trimethygallium (TMG) and triethylgallium
(TEG) in hydrogen and nitrogen atmospheres was studied by a
quadrupole mass analyser. The decomposition reactions of TMG in H_2
and in N_2 and of TEG in H_2 and in N_2 take place in the temperature
ranges 370-460°C, 450-570°C, 220-330°C and 270-380°C respectively.
The reaction mechanisms are hydrogenolysis for TMG in H_2, homolytic
fission for TMG in N_2 and β elimination for TEG in both H_2 and N_2.
The hydrogenolysis of TMG follows homogeneous first-order kinetics
on TMG concentration.

J.Electrochem.Soc. 1985, 132:677. Reprinted by permission of the
publishers, The Electrochemical Society Inc., Pennington, NJ, USA.

Zanio, K.

THE EFFECT OF INTERDIFFUSION ON THE SHAPE OF HgTe/CdTe SUPERLATTICES

The shape and widths of HgTe wells in HgTe/CdTe superlattices are calculated with a growth and interdiffusion model. For a growth temperature of about 200°C the model shows the HgTe wells to expand up to about an angstrom and CdTe tails to penetrate several angstroms into the HgTe wells. At superlattice growth temperatures, the extent of interdiffusion from model profiles agrees qualitatively with experimental results.

Reprinted with permission from J.Cryst.Growth 1986, $\underline{A4}$(4): 21 06 published by American Institute of Physics, © 1986, American Vacuum Society.

Zulehner, W.

CZOCHRALSKI GROWTH OF SILICON

The importance of Czochralski (CZ) grown silicon as a basic material for solid-state electronics is outlined. After that, a short review is given of the different possibilities for growing silicon crystals and the history of Czochralski technique. Details of the CZ pulling procedure are discussed. The different aspects of the process like starting material, crucible problems, heat and melt flow conditions, interface reactions, as well as incorporation and distribution of impurities in the crystals are considered. Examples are given of typical and exceptional impurity distributions. Oxygen is of supreme importance in Cz-Si and is dealt with in greater detail. Its origin, typical concentrations and distributions and its influence on crystal quality are described. Finally, the technical aspects of current CZ-Si growth are outlined and the mechanical designs of the pulling apparatus, including recharging equipment, are illustrated.

Reprinted with permission from J.Cryst.Growth 1983, $\underline{65}$:189, published by North Holland Publishing Co., Amsterdam.